'Is new technology a blessing or and powerful book, Bob Hughes sl that the answer depends, above all, on whether society is plutocratic or egalitarian. A must-read for all who care about humankind's future.'

James K Boyce, University of Massachusetts Amherst, US.

'Bob Hughes blows all the hype out of the water. He understands both the history and the technology. In this wonderful book he explains how inequality turns humanity into a destructive force that then uses technologies to cause harm.'

Danny Dorling, Professor of Geography at the University of Oxford.

'Hughes' book is much needed in helping us get beyond the glib national policies that seem to take both technology and inequality as givens. *The Bleeding Edge* deconstructs the causes of inequality and, while helping us understand that technology is not a panacea, it lets us truly understand for whom technologies are developed and sold.'

Joan Greenbaum, Professor Emerita of Environmental Psychology, City University of New York, US.

'A fascinating study of how inequality inhibits technological innovations and how only an egalitarian society can truly sustain progress.'

Hsiao-Hung Pai, author of *Chinese Whispers, Scattered Sand, Invisible* and *Angry White People*.

'The *Bleeding Edge* is truly the leading edge of books that challenge us to rethink the relationship between technology, capitalism and inequality. Rejecting both apocalyptic pessimism and techno-optimism, Hughes provides a compelling map to the future in which information technologies are harnessed for the common good. Powerfully argued and easy to read, this is one of those books that can help change the world.'

Betsy Hartmann, Professor Emerita of Development Studies and senior policy analyst, Population and Development Program, Hampshire College, US.

'Hughes nails inequality to the wall with precision and passion. He weaves together multiple strands to make the case against inequality, from economics through anthropology; from evolutionary theory through social epidemiology. Then, once he has constructed his airtight logic, he colors it in with the emotional dimensions of life lived under oppressive hierarchy – or empowering egalitarianism. Hughes' book dares us to stop begging for half-measures and instead demand our human birthright: full social and economic equality!'

Deborah S Rogers, President of Initiative for Equality, and affiliated to the Institute for Research in the Social Sciences at Stanford University, US.

'Technology comes between us and our environment but, in this highly original book, Hughes shows how inequality comes between us and our technology. Inequality subverts technical progress, increases its environmental damage and prevents it from satisfying our real needs. This is a thesis we cannot afford to ignore.'

Richard Wilkinson, Professor Emeritus of Social Epidemiology at Nottingham University, UK, and co-author of *The Spirit Level*.

THE BLEEDING EDGE

About the author

Bob Hughes worked as a schoolteacher, calligrapher, and in advertising before getting involved with computers in the mid-1980s, working on interactive information systems, running an interest group, and writing about the new industry's unofficial history and creative traditions. Later, he became involved in campaigning for the rights of migrants, on whose labor the digital economy is built, and was a co-founder, in 2003, of No One Is Illegal UK. He taught digital media at Oxford Brookes University till 2013, with a particular interest in publishing for social change. He now lives and writes in southern France.

Acknowledgements

This book exists thanks to the kind people who read drafts, made suggestions, and encouraged it on its way. They include: Danny Dorling, Ursula Huws, Chris McEvoy, Nick Nuttgens, Hsiao-Hung Pai, Natasha Stotesbury, Joe Shaw, Ciaran Walsh, Betsy Hartmann and Jim Boyce, Nancy Lindisfarne and Jonathan Neale, Sarah and Bob LeVine; and my friends at Oxford Brookes, especially Tom Betteridge, John LoBreglio, and all my colleagues in the Centre for Publishing Studies. I particularly want to thank Jonathan Rosenhead (emeritus professor of Operational Research at LSE), and Raul Espejo (Syncho Research, Lincoln; former leader of Chile's Cybersyn project) for their time and patient explanations, which I hope I've not travestied.

My biggest debts are to Teresa Hayter, who lived with, read and tactfully critiqued the book as it evolved; and to New Internationalist for taking this on so wholeheartedly and so well – and for being such a clear and consistent voice for sanity and justice through more than 40 years of planetary mayhem.

THE BLEEDING EDGE
Why technology turns toxic in an unequal world

Bob Hughes

The Bleeding Edge
Why technology turns toxic in an unequal world

First published in 2016 by
New Internationalist Publications Ltd
The Old Music Hall
106-108 Cowley Road
Oxford OX4 1JE, UK
newint.org

© Bob Hughes

The right of Bob Hughes to be identified as the author of this work has been asserted in accordance with the Copyright, Designs and Patents Act 1998.

All rights reserved. No part of this book may be reproduced, stored in a retrieval system or transmitted, in any form or by any means, electronic, electrostatic, magnetic tape, mechanical, photocopying, recording or otherwise, without prior permission in writing of the Publisher.

Edited by Chris Brazier
Designed by Juha Sorsa
Front cover adapted by Juha Sorsa from Gustave Doré's 'Moses breaking the tablets of the law', 1866
Indexed by Angie Hipkin

Printed by T J International Limited, Cornwall, UK
who hold environmental accreditation ISO 14001.

British Library Cataloguing-in-Publication Data
A catalogue record for this book is available from the British Library.

Library of Congress Cataloging-in-Publication Data.
A catalog record for this book is available from the Library of Congress.

ISBN 978-1-78026-329-8
(ISBN ebook 978-1-78026-339-7)

Contents

Foreword by Danny Dorling   6

Introduction   9

1 Technofatalism and the future – is a world without Foxconn even possible?   15
Two paradoxes about new technology
Humanity began with technology
Technology emerges from egalitarian knowledge economies
The myth of creative competition
Why capitalism inhibits innovation
Capitalism didn't make computers... but took computing down the wrong path

2 From water mills to iPhones: why technology and inequality do not mix   46
Egalitarian hopes for computing
The return of medieval economics
The first modern environmental crisis
An unequal society is a dangerous place for powerful ideas
Water mills, and how new technology can be a curse
Firearms take a European turn

3 What inequality does to people   73
Inequality reduces life expectancy
Equality and the Soviet Union
Autonomy and solidarity: the essential nutrients
Inequality makes people shorter
Today's inequality will damage future generations

4 The environmental cost of human inequality   91
Are the rich destroying the earth?
Inequality turns humans into a geological force
Malthus's mistake: not too many babies, but too much debt
Ehrlich's last gasp: technology and 'eye-pat'
The power to choose a low-impact life

THE BLEEDING EDGE

5 Ever greater impact, ever less benefit: high-tech capital's mysterious lack of growth   111
'Keep your nerve' or 'tough it out'
Why computers have grown nothing but themselves
Inequality: the elephant in the room

6 The invisible foot: why inequality increases impact   123
Technology plus inequality equals meltdown
'Positionality' and 'human nature'
Traffic waves and why faster is slower
Computers and the positional economy: obsolescence gone mad
The rise of financial services, trailed by women in old cars
Putting a girl on the moon: the cost of education
How 'e-learning' rebounded on the poor

7 Enclosure in the computer age: the magic of control   146
The supernatural enters everyday life: the magic of commodities
Power over theW future: the magic of intellectual property
Computers and the making of money
The world gets smaller and hotter
Closing the technological frontier (or trying to)
Other routines are possible!

8 Sales effort: from the automobile to the microchip   170
The all-steel automobile as an energy sump
How the sales effort shaped the chip
Moore's self-fulfilling prophecy: chips with everything
Dictating the future
The visionary turn
Embracing carnage: faith in disruption

9 Technoptimism hits the buffers   190
The toxic deWmands of purity
Obsolescence and e-waste: a total system
Displacing the problem to Africa
Entropy: measuring what's possible
Maxwell's demon: the spoiler in the green growth dream
Puncturing the weightless economists

10 The data explosion: how the cloud became a juggernaut   210
Forced migration: corporate flight into the cloud
How the web became an entropy pump
The cost of the dotcom bubble and Web 2.0

11 'The least efficient machine humans have ever built': how capitalism drove the computer down a dead end   230
    The buried world of analog computing
    Clocks: why today's computers mostly do nothing, but very quickly
    Soviet computing: diversity under scarcity and bureaucracy
    Time-sharing: another abandoned road
    Competitive pressure narrows all options

12 Planning by whom and for what? The battle for control from the Soviet Union to Walmart   251
    The benefits and dangers of centralized planning
    Electrification of the Soviet Union: heteronomous planning becomes the global norm
    Linear programming, with and without computers
    The curious incident of the capitalist calculation debate
    Connection-making and the ecology movement
    Operational Research and cybernetics
    Variety engineering: the difference between amplification and shouting

13 A socialist computer: Chile, 1970-1973   280
    A global crisis of inequality
    The Unidad Popular: a moderately egalitarian program
    Stafford Beer and 'cybernetic socialism'
    How much computer hardware does a viable society need?
    Cheap, radical technology
    'War' is declared

14 Utopia or bust   303
    Envisioning Utopia: the world turned right way up
    Utopian practicalities: food and work
    Beauty and lower impact, from the bottom up
    Shrinking roads, expanding diversity
    Putting babies and children at the heart of the economy
    Shared work: Utopia's powerhouses
    Community is stronger than we think: 'Disaster Utopias'
    The Right knows the power of solidarity, even if the Left doesn't
    Equality, truth and the experience of being believed
    The 'apparatus of justification'

Foreword

Technology is neutral. It is the adaptation of knowledge to practical purpose. How we use technology and what technologies we develop is up to us. The machine is not in control, corporations and politicians are.

Artificial intelligence is again all the rage. But it has not been artificial intelligence that has made our world more unequal. It has been us, and only in those countries which chose the road to greater inequality, to a new serfdom, from the 1970s onwards.

Less able economists list 'technology' as one excuse for growing economic inequality. They are unaware that the most technologically advanced country in the 1980s, Japan, was (and is) also one of the world's most economically equal countries – and it still produces the highest number of inventions (patents) per head today. Those economists are also unaware that equitable Scandinavian countries are responsible for the greatest production of scientific papers per head. A few don't even know the origins of the word *entrepreneurial*, or just how equitable France is compared with the UK and US, and how much better off the average French family is as a result, and how early and widely the French adopted Minitel, the precursor to the World Wide Web. What the most unequal countries of the world are best at is advertising, which creates the impression of being innovative.

Bob Hughes blows all the hype out of the water. He understands both the history and the technology. In this wonderful book he explains how inequality turns humanity into a destructive force that then uses technologies to cause harm.

More egalitarian societies, on the other hand, tend to use it for better ends. It is in more egalitarian societies that technologies are used to preserve and increase biodiversity. In these societies most children since the 1990s have been taught on the assumption that they had great ability and then grew up to be able to use technology well. I recently gave a talk to several hundred civil servants in London, many of whom will have gained top-level degrees from prestigious universities and passed fast-track examinations. None of them could fix their bespoke computer system when the slides would not show. A young man in a 'service role' had to be summoned to do that. He appeared to be the only black person in the room. Inequality also creates and sustains racial inequalities. It is neither technology nor ability that makes us become more unequal.

Economic inequality has no 'safe level', but high inequality is far more harmful than low inequality. Bob Hughes suggests that we should treat high economic inequality as we treat a banned substance that we know to be harmful. He concludes that encouraging inequality, or tolerating it, when you have the power to do something about it, should be seen as a crime against humanity – given the harm that results. Once you consider the harm that is being caused by the misuse of technology, from drone strikes killing people to subversive marketing destroying the meaning of lives and reducing humans to consumers, it is hard not to have great sympathy with his position.

Equality, not competition or hierarchy, is 'the mother of invention'. Progress slows or goes into reverse as soon as new ideas are appropriated by hierarchical organizations. Inequality is sustained when a few people persuade the majority that there is no alternative. But there are always huge numbers of alternatives – the computer's history is littered with them, and the 'next big thing' always comes out of one of the alternatives. It is then that corporations in the world's most unequal rich countries seek to appropriate the new idea and deter further real innovation, as they did to great effect in the 1990s. They will seek instead to sell you a phone where the battery life will rapidly decrease, the operating system updates will overpower its resources, and you will, before you know it, be forced to buy their next product – which does little more than the last. Profit maximization is the anathema of true innovation.

An unequal society can only stay unequal by systematically and continuously preventing people from working together and helping each other – and often criminalizing their attempts to do so. In the UK and the US our imaginations in the 2000s became cramped by generations of rising inequality and the propaganda justifying it. What we need now is to start understanding how different, more exciting and richer our lives will be once great inequality has ceased to be seen as a legitimate state of affairs.

All societies that become greatly unequal eventually fail to adapt technologically and become unsustainable. The water stops flowing so well in the aqueducts. The crops fail because the soil becomes exhausted. The roads and bridges wear away and you don't create new technologies to replace them. People in far-flung places start to ask 'what did the Romans ever do for us?' and begin to realize that what they did, they did a long time ago, often through appropriating and monopolizing technologies that others invented. When inequality rises, you pour more and more resources into your armies and into new technology for them – not into truly productive and co-operative purposes. Slaves and scribes are left to do the hard work. Eventually the empire falls – because high and rising inequality and a consequent lack of real innovation and adaptation is how it so often ends.

We can now see this end coming. It is not too late to learn and change. And technology can help us do this if we better understand it – and ourselves.

Danny Dorling
Halford Mackinder Professor of Geography, Oxford University

Introduction

In northern California in 1974, it was hobbyists, draft-dodgers and political activists who cobbled together what they impiously called 'personal computers' from mail-order, bootleg and 'liberated' components. Few of them had any intention of founding a new capitalist industry and many of them explicitly opposed any such idea. Some of them wanted a political revolution, and a few recognized the danger that a counter-revolution would sneak into the new movement before it had fairly got started, and end up making the world worse instead of better.

Two events of that period are iconic. First, in 1974, Ted Nelson, a social scientist and literary scholar, self-published a book called *Computer Lib*, with a clenched fist on its cover, and the slogan 'You can and must understand computers, NOW!' It was never a big seller (thanks to Nelson's eccentric approach to publishing) but it became a sort of foundational text for what would become the 'hacker' movement, which subsequently produced such things as the famous GNU/Linux computer operating system. *Computer Lib* explained basically how computers worked, some of the different things they were capable of, and how they can either confuse, stupefy and oppress, or enlighten, empower and liberate. The book is peppered with memorable insights and slogans; for example: 'The purpose of computers is human freedom' and (in an updated 1987 edition): 'In 1974, computers were oppressive devices in far-off air-conditioned places. Now you can be oppressed by computers in your own living room.' And this (from one of Nelson's websites) about the myth of 'Technology':

A *frying-pan* is technology. All human artifacts are technology. But beware anybody who uses this term. Like 'maturity' and 'reality' and 'progress', the word 'technology' has an agenda for your behavior: usually what is being referred to as 'technology' is something that somebody wants you to submit to. 'Technology' often *implicitly refers to something you are expected to turn over to 'the guys who understand it'.*

Nelson, who became best known as the originator of the concept of 'hypertext' (which we all use nowadays, after a fashion, when we use the World Wide Web) has been described as the computer underground's Tom Paine. The idea that you might 'understand computers, NOW!' (rather than just buy one, and watch TV on it) may now sound quaint, but the common belief that we will never be able to understand them (and a dominant culture that discourages you from trying) has certainly not helped the cause of human freedom. It has also helped to build the myth that what we have now got is the best of all possible worlds and we shouldn't even try to imagine anything better.

The second iconic event, in 1975, was the 20-year-old Bill Gates's angry challenge to the 'thieves' at San Francisco's Homebrew Computer Club who had copied and distributed his version of the BASIC computer language without paying for it, and their own outrage that Gates expected them to pay. Club members and their friends had, after all, just created a less comprehensive but still highly capable version of the language (Tiny BASIC) which anybody could have for nothing. Individuals and groups had poured effort into it for the sheer pleasure of taking on an impossible-looking challenge and pulling it off in style. Not only did they want nothing for it; they did not particularly care which team or individual finally cracked the problem. It was good enough that somebody had cracked it. To claim proprietorship of such a thing seemed obscene.

Twenty years later, in 1995, Gates's Microsoft Corporation was becoming a global economic force and its monopolistic tendencies were the subject of a US government investigation – only the fourth company in US history to have merited that kind of intervention. The distinguished science writer James Gleick pointed out that here, for the first time ever, was a major company that 'does not control a manufacturing industry (as IBM did), a natural resource (as Standard Oil did) or a regulated public utility (as AT&T did).' Instead, by strenuous assertion of legal rights and precedents, it had come to own 'the standards and architectures that control the design of modern software'. Gleick made it very clear that Microsoft seriously intended at the time to corral as much of the world's knowledge as it could get away with, and extract astronomical rental income from it. At the time, there was a widespread sense of amazement that a business could even attempt such a thing, and achieve so much power simply from 'owning' knowledge.

Another two decades on, the idea that great fortunes are built on intellectual property has become totally normalized and uncontroversial, and is even enshrined in international trade treaties. Business empires cross fresh social and personal boundaries as routinely as they do official, international ones – and our technologies help them do it.

Each time, somehow, we adapt swiftly to the 'new normal'. But a point comes when we no longer have the right to give way so gracefully.

I argue in this book that escalating human impact on the earth has gone hand in hand with successful encroachments on egalitarian culture, as with the neoliberal onslaught since the 1970s but extending far back in history. The issues go far beyond computers and electronics, intellectual property law, or even modern global capitalism. To get to the root of the matter we will need to wind the tape back to where it all began, when mercantile elites first acquired 'the right' (because they made the laws) to own whatever they needed to own and to disown anything and anyone that might be a liability; to the time when we became 'modern' inasmuch as we learned to stand by while others starve, and to tolerate and even to respect those who take more than their share.

We are involved in the endgame of something that began in the squalor of medieval Europe. The challenge cannot be resolved until we tackle the social failure that set it in motion: entrenched inequality, and the genteel acceptance of it.

Bob Hughes
St André de Rosans, France, April 2016

1

Technofatalism and the future: is a world without Foxconn even possible?

The routine assumption is that progress – particularly high-tech progress – depends on inequality, on a world of capitalist entrepreneurs, low-paid factory workers and toxic waste dumps. Yet every major development in the computer's history arose from voluntary initiative or public funding rather than corporate research. The historical evidence suggests that innovation and creativity thrive in egalitarian settings and are stifled by competition. Far from deserving credit for the computer revolution, capitalism has driven it down a narrow and barren path, and might even have turned it into 'a revolution that didn't happen'.

In May 2010 world media picked up a report from the Hong Kong-based *China Labour Bulletin* that desperate workers were killing themselves by throwing themselves out of the windows of the vast Foxconn factory in Shenzhen, in China's Guangdong province, where the Apple iPhone was being manufactured.[1] Newspaper columnist Nick Cohen wondered what could be done to alleviate the situation, or even to stimulate some sense of outrage about it, but drew a blank:

> A boycott of Foxconn's products would not just mean boycotting Apple, but Nintendo, Nokia, Sony, HP and Dell too. Boycott China and you boycott the computer age, which, despite the crash, effectively means boycotting the 21st century, as we so far understand it.[2]

Cohen's 'as we so far understand it' does at least hint at a recognition that 'another world is possible', but he did not pursue the idea. The phrase 'a quick trip back to the Stone Age' seems to lurk not far away.

It's drummed into us that all good things come at a price. If we want nice things, someone must pay for them: you can't have modern, high-tech luxuries *and* happy workers, clean rivers, lovely woodlands and country lanes thick with butterflies in summer.

It seems impossible to live what we think of as 'a normal life' in what we've learned to call 'a modern country' without being complicit in human immiseration or environmental destruction. Not even the poor can avoid complicity; in fact, they least of all – from the 19th-century factory-hands in their slave-grown cotton clothes, sustained by Indian tea sweetened with slave-grown sugar, to the 21st-century migrants whose very existence can depend on having a mobile phone. 'Progress' apparently requires inequality.

The range of a modern economy's inequality is astonishing and all of it is packed into its most popular products. As technology advances so does the range of the inequality that's drawn into its web. Must it be so? Today's iconic electronic products, like yesterday's cotton ones, embody the greatest range of human inequality currently possible. Most of us know at least some of the facts: the toxic waste mountains; the wholesale pollution of the environments where the copper, gold, tin and rare-earths are extracted, in countries where life itself has never been so cheap; the sweated labor; and so on.

TWO PARADOXES ABOUT NEW TECHNOLOGY

Yet here's the first of two key paradoxes: when you look at what actually happens when technological progress is made, you find very little to support the idea that progress demands inequality – and even some mainstream economists recognize this. World Bank economist Branko Milanovic, for example, concluded a large-scale study of inequality and economic growth in history like this:

The frequent claim that inequality promotes accumulation and growth does not get much support from history. On the contrary, great economic inequality has always been correlated with extreme concentration of political power, and that power has always been used to widen the income gaps through rent-seeking and rent-keeping, forces that demonstrably retard economic growth.[3]

This is especially and manifestly true when looking at the present system's 'jewel in the crown': the computer. The thing we know (or think we know) as 'the computer' emerged in conspicuously egalitarian settings, and it wouldn't go on functioning for very long if inequality ever succeeded in its quest to invade every nook and cranny of the industries that support it.

The computer in your hand may have arrived there via a shocking toboggan-ride down all the social gradients known to humanity, but inequality was conspicuously absent at its birth, largely absent during its development, and remains alien to computer culture – so alien that the modern economy has had to create large and expensive 'egalitarian reservations' where the essential work of keeping the show on the road can be done in a reasonably harmonious and effective manner. *The New Yorker*'s George Packer has described[4] how the leading capitalist companies (Google, Microsoft, Apple and the like) have even built their own, luxurious, egalitarian 'villages' and 'campuses' where their programmers and other creative types are almost totally insulated from the extreme inequality around them, and can believe they have moved beyond capitalism into a new egalitarian age.

More than half of the world's computers and smartphones, more and more of its electronic appliances, and nearly all of the internet, depend on software created by freely associating individuals, in conscious defiance of the management hierarchies and profit-driven, intellectual property (IP) regime that underpin giants like Apple. Richard Stallman, founder of the 'Free Software' movement, sees any attempt to take ownership of the process as an affront to humanity. Of intellectual property law, Stallman has said:

THE BLEEDING EDGE

> *I consider that immoral... and I'm working to put an end to that way of life, because it's a way of life nobody should be part of.*[5]

Free-market hawks may sneer at such idealism but their world would simply not exist without people like Stallman. Even if it did, it would not work very well without Stallman's brainchild, the computer operating system known as GNU/Linux, and the global network of unpaid collaborators who have developed and continue to develop it. Google itself is built on GNU/Linux, as are Facebook and other social-media sites, and even the computers of the New York Stock Exchange: GNU/Linux is faster and more robust than the commercial alternatives.

Stallman wrote and published GNU in 1983 as an alternative to the older, established Unix operating system,[6] with the difference that all of the code was freely available to anyone who wanted it, and could be changed and improved by anyone capable of doing so (hence 'free and open source', or FOSS). Stallman's only stipulation was that nobody could own the code, and any modifications must be shared. GNU became 'GNU/Linux' after 1991, when a teenage fan of Stallman's work, Linus Torvalds, started to circulate the code for the 'kernel' that allows GNU to run on different kinds of computers.[7] This made it possible to use GNU/Linux (now generally known simply as Linux) on just about every kind of computing device that exists, including automobile engines, avionics, industrial appliances, power stations, traffic systems and household appliances.

The second paradox is that, while the new technologies are in principle supremely parsimonious, their environmental impact has turned out to be the exact opposite. Each wave of innovation needs fewer material inputs than its predecessor to do the same amount of work – yet in practice it consumes more resources. The Victorian economist William Stanley Jevons (see Chapter 5) was the first to draw attention to this paradox, which now bears his name – and it becomes even more striking when considering all the industries and activities that depend on or are mediated by electronics and computers. As economic activity has been computerized, it has become more centralized, and its overall environmental impact has increased – as have control by capital of labor and of people, the wealth-differences between rich and poor, and the physical distances between them.

Is this mounting impact an inevitable 'price of progress', or is it the result of progress falling into the hands of people and a system that simply cannot deal with it responsibly?

WHAT IS TECHNOLOGY ANYWAY?

It is important to challenge two conventional assumptions that are often made about technology: first, that we have capitalism to thank for it; and second, that it follows a predetermined course, that the future is waiting to be revealed by clever minds and that progress 'unfolds' from Stephenson's Rocket to the automobile, DVDs and the iPhone.

The economist Brian Arthur, who has made a lifetime study of technological change, argues that human technology is a true, evolutionary phenomenon in the sense that, like life, it exploits an ever-widening range of natural phenomena with ever-increasing efficiency: hydraulics, mechanical, electrical phenomena and so on. He defines technology as:

> *a phenomenon captured and put to use. Or more usually, a set of phenomena captured and put to use... A technology is a programming of phenomena to our purposes.*[8]

Technology develops through greater and greater understanding of the phenomena, and what they can be made to do, and how they can be coaxed into working together. Arthur uses the analogy of mining: easily accessed phenomena are exploited first (friction, levers) then 'deeper', less accessible ones (like chemical and electrical phenomena). As understanding of the phenomena deepens, their essential features are identified for more precise exploitation: the process is refined so that more can be done with less.

As the 'mining of nature' proceeds, what once seemed unrelated ventures unexpectedly break through into each other's domains, and link up (as when magnetism and electricity were discovered to be the same phenomenon in the late 18th century). No technology is primitive; all of it requires bodies of theory, skill and experience; it tends inexorably to greater and greater economy of material means. He describes how phenomena – for example, friction being used to make fire – are exploited with increasing efficiency as they are worked with, played with and understood.

The parallels with biology are striking. Technology is just like a biological process – and there is a tendency at this point (which Arthur goes along with somewhat) to start thinking of technology as 'a new thing under the sun' with a life of its own, and rhapsodizing about 'where it is taking us'.

If you only look at the technologies themselves, in isolation, the parallels are there, including the tendency to see computer code as space-age DNA, and to sit back and be awed as some brave new world unfolds. But what really distinguishes human techology from biological evolution, surely, is that it all happens under conscious, human control – which implies some important differences.

Technologies, unlike living organisms, can inherit acquired traits, and features of unrelated technologies can, as it were, 'jump species', as when turbine technology migrated from power stations into jet engines, and punched-card technology for storing information spread from the textile industry (the Jacquard loom) into the music industry (the pianola) and then to computing. The eclectic human agency responsible for this cross-fertilization is well demonstrated by the Victorian computer pioneer Charles Babbage, who was continually investigating the arcane processes developed in different industries – and made the connection with the Jacquard loom at an exhibition in 1842, as did his future collaborator, Ada Lovelace.[9]

This is even more the case where electronics and computers are concerned – a point that Brian Arthur makes: 'Digitization allows functionalities to be combined even if they come from different domains, because once they enter the digital domain they become objects of the same type – data strings – that can therefore be acted upon in the same way'.[10] Digitization is, moreover, just one of the possible techniques for doing this, as will be explained later.

The underlying and really powerful principle is that phenomena from utterly different domains of experience may share a deeper, abstract reality that can now be worked with as if it were a physical thing in itself.

Most importantly of all, technological evolution need never have dead ends – and this is where we come slap-bang up against the contradiction that is today's technological environment, in which promising technologies can be ditched, apparently never to return, within months of their first appearance.

TECHNOLOGY SHOULD HAVE NO DEAD ENDS

In principle – and in practice for most of the millennia that our technological species has existed – ideas that have 'had their day' are not dead and buried for ever. Human culture normally sees to that. Technological improvements can and should be permanent gains – inventions should stay invented. They may lurk in human culture for decades or even centuries, and be resurrected to become the bases of yet more discoveries, so that technology becomes richer, more complex and more efficient.

In the past, discoveries have tended overwhelmingly to become general property, rapidly, via exactly the same irrepressible social process whereby songs and jokes become general property. The genie does not always go back into the bottle and can turn up anywhere – precipitating further discoveries, always making more and yet more efficient use of natural phenomena, and revealing more *about* those phenomena, which yet more technologies can then use.

Biological evolution proceeds blindly, as it must, over vast epochs via small changes and sudden catastrophes. It contains prodigious numbers of dead ends: species that die out for ever, taking all their hard-won adaptations with them. Living species cannot borrow from each other: mammals could not adopt the excellent eyes developed (in the octopus) by molluscs; we had to develop our own eyes from scratch; so did the insects. Human technologies can and do borrow freely from each other, and in principle have no dead ends.

Unlike biological species, an abandoned technology can lie dormant for centuries and be resuscitated rapidly when conditions are right. With living things, there is no going back; the fossilized remains of extinct species, like ichthyosaurs and pterodactyls, can't be resuscitated when the climate is favorable again. Darwinian evolution must plough forward, the only direction available to it, and create completely new creatures (dolphins, birds) based on the currently available stock of life forms (mammals, reptiles). But with technology we can always go back if we want to. For once, the arrow of time is under our control. Or should be.

Comparing Darwinian and technological evolution reveals an anomaly in the kind of innovation we see around us in the present computer age: here, technologies apparently *can* effectively disappear from the common pool, the way dinosaurs and other extinct species have done. Fairly large technologies can disappear abruptly, as soon as a feeling spreads among those who control their manufacture that the market for them might soon disappear, or even might become less attractive.

Or a technology may deliberately be kept out of the common pool, by someone who patents it in order to suppress it. Yesterday's ideas may survive in documents, and for a while in human knowledge and skill, but they soon become very difficult to revive. 'The show moves on.' Premises and equipment are sold, staff are laid off and all the knowledge they had is dispersed; investors pull out and put their cash elsewhere; and products that once used the technology either die with it, or are laboriously redesigned to use alternatives. These extinctions help to create the determinist illusion that technology follows a single 'best' path into the future but, when you look at what caused these extinctions, fitness for purpose seldom has much to do with it.

No technology ought ever to die out in the way living organisms have done. It seems perverse to find Darwinian discipline not merely reasserted in a brand-new domain that should in principle be free of it, but in a turbo-charged form, unmitigated by the generous time-scales of Darwinian evolution. This market-Darwinism comes at us full pelt within ultra-compressed, brief, human time-frames. Where there should be endless choice, there is instead a march of progress that seems to have the same deterministic power as an avalanche.

But this is a fake avalanche. Every particle of it is guided by human decisions to go or not to go with the flow. These are avalanches that can be 'talked back up hill' – in theory and sometimes even in practice. Even in the absence of such an apparent miracle, deviation always remains an option, and is exercised constantly by the builders of technology. Indeed the market would have very little technological progress to play with if technologists did not continually evade its discipline, cross boundaries, and revisit technologies long ago pronounced dead. This becomes more and more self-evident, the more our technologies advance.

ARE SOCIETIES TECHNOLOGIES?

Brian Arthur begins to speculate on the possible range of things that might be called 'technology'. He observes that science and technology are normally paired together, with science generally assumed to be technology's precursor, or its respectable older brother. Yet he points out that human technology evolved to a very high level for centuries and even millennia before science existed, as we now understand it. And then he asks, is modern science a technology? It is a technique that, once discovered, has evolved in much the same way as specific technologies have done.

Taking this argument further, human nature is part of nature; we have various ways of exploiting it to particular purposes and, as we learn more about how people function, those ways become more and more refined.

Exploiters of humanity are avid students of human nature: they are eagle-eyed at spotting ways of coercing people to do things they do not wish to do and quick to adopt the latest research for purposes of persuasion. They know that human nature is what we make it. They make it fearful and obedient. We, however, know that human nature can be better than this. We know that human nature can take almost any form – but we also know, roughly at least, what kind of human nature we want. Should we not devise societies that will help us to be the kinds of people we aspire to be?

THE BLEEDING EDGE

A key part of any Utopian project should be to discuss widely and think deeply about the human natures we want to have and the ones we do not want to have, and to devise the kinds of social arrangements that will support and reward those characteristics.

HUMANITY BEGAN WITH TECHNOLOGY

Economic policy is driven by an assumption that technology is something hard, shiny and baffling that emerged in the cut-and-thrust of late 18th-century northern Europe, and has since spread throughout the world from there, bringing a mix of great benefits and serious challenges that we take to be an inevitable concomitant of progress. It's further assumed that the vehicle for this revolution was the capitalist company.

Taking Brian Arthur's definition of technology as 'a phenomenon captured and put to use', it's pretty clear that technology is a lot bigger than that, and a lot older than that. It's now becoming apparent that the people of so-called 'primitive societies' were and are great and pioneering technologists – and none of today's technologies would be conceivable without what they achieved (so the 'giants' whose assistance the great Isaac Newton modestly acknowledged were themselves 'standing on the shoulders of giants': the Human Pyramid itself).

Richard Rudgley, an anthropologist, has described the scale of these discoveries in a book published in 1998, *Lost Civilisations of the Stone Age*.[11] Long before the first cities appeared, leaving their large and durable remains for the first archeologists to ponder over, humans in all parts of the world were developing highly efficient tools and techniques for making tools, had elaborate cuisines, were great explorers and expert navigators, artists and students of the natural world, including the sky. They even practiced surgery. We know this because evidence has been found in prehistoric remains from all over the world, of the challenging form of cranial surgery known as trepanning (to relieve pressure on the brain caused by blood clots); one of the few forms of surgery that leaves unambiguous skeletal evidence. It is reasonable to assume from this that they also knew many other kinds of surgery.

Martin Jones, a pioneer of the new techniques of molecular archeology, makes the point that humans are not even viable without at least minimal technology, such as fire. In his book *Feast: Why Humans Share Food*, Jones says that 'human evolution may have something to do with reducing the costs of digestion'.[12] Humans have relatively small teeth and jaws, and our guts are not long enough to cope well with a diet composed entirely of uncooked food. Cooking also neutralizes the toxins in many otherwise inedible plants, increasing the range of foods humans can use. All of this requires highly co-operative sociality – which is in turn facilitated by the large, anthropoid brain that became possible through reduced 'metabolic expenditure' on jaws and guts: a self-reinforcing feedback cycle that, at a certain point, produced the intensely sociable, essentially technological, highly successful human species. Humans, their technology and their distinctive social order all seem to appear simultaneously in the archeological record 100,000 or more years ago.

TECHNOLOGY EMERGES FROM EGALITARIAN KNOWLEDGE ECONOMIES

Throughout nearly all of their first 100,000 or so years, the dominant characteristic of human communities has been egalitarianism, and we can work out a lot about how these egalitarian societies functioned not only from the physical evidence they have left, but also from modern people who live radically egalitarian lives: today's hunter-gatherer and foraging peoples. Many of these communities have brought the art of egalitarian living to a level of impressive perfection, and have independently developed many of the same social mechanisms for maintaining equality – particularly significant because they are so widely separated from each other, on the furthest and least-accessible margins of all the inhabited continents in the world. One of these characteristics, which almost everyone who meets them comments upon, is an unshakeable commitment to sharing knowledge. To borrow a useful phrase, they are the ultimate 'knowledge economies'.

But there is much more to this than 'sitting around all day

talking', which is what so many Europeans see when they come across indigenous communities. There is an extraordinary commitment to accuracy and truth. Hugh Brody – an anthropologist who has worked on land-rights campaigns with hunter-gatherer communities, and made documentaries with them – has reflected on this at some length in his book *The Other Side of Eden*.[13] George Dyson, whose work on computer history will be mentioned later, has also written about the extraordinary technological traditions this kind of knowledge economy can support, in his book about the Aleuts and their kayaks, *Baidarka*.[14] Aleut kayaks are made in some of the most resource-poor places on earth, and are technological miracles that defy long-accepted wisdom by travelling at speeds once considered theoretically impossible for a human-powered craft.

The hunter-gatherer knowledge economy also supports a healthier kind of person. Physically, hunter-gatherers have always been healthier and often taller than their civilized counterparts (see Chapter 3). Explorers and anthropologists constantly remark on their happiness and 'robust mental health'. Brody attributes this to a complete absence of anxiety about being believed, or listened to, or being completely honest, or whether the other person is telling the truth. This has a utilitarian dimension – such societies simply cannot afford deceit and lives depend on absolutely accurate information – but it runs deep: this is how we evolved. Evolution made us radically honest people, and going against this hurts.

Wherever it is found, the egalitarian ethos is maintained through what another anthropologist, Christopher Boehm, identified as 'counter-dominance' strategies.[15] We can readily recognize these at work everywhere in modern communities in the extensive repertoire of strategies for 'taking someone down a peg or two', ranging from friendly ribbing, to gossip, to ostracism and, in the extreme, to homicide. There is also the array of self-effacement strategies used by those who do not want to seem domineering: 'honestly, it was nothing'; 'I'm completely hopeless with computers', etc. Even within the most hierarchical and unequal modern societies, personal life is lived as much as possible within egalitarian or would-be egalitarian social bubbles (families, peer groups, work-groups, neighbors and, in wartime and warlike situations, nations).

In fact, we seem to need these even more as societies become harsher and more stratified, and it is now gradually becoming recognized that the evils that arise from inequality are largely the effects of *group* inequality – 'us' against 'them'.[16] We gravitate towards groups where we can have this experience of solidarity and, what is more, we do it without being aware that we are doing so. This is why evil is so banal; why ordinary people who see themselves as decent folk (and are, in most situations) are capable of genocide.

Solidarity is a fundamental phenomenon of human nature – and dominant forces have learned down the centuries to exploit it. If technology is 'a phenomenon captured and put to use' then all our formal and informal social systems are some kind of technology, and 'social engineering' is what they do. We need social systems that maximize our chances of 'not doing evil', to borrow Google's motto – which is precisely what Google's practice of segregating its creative elite in pretend-Utopias, separate from the society around them, can't possibly do.[17]

Theologian-turned-neuroscientist Heidi Ravven has documented the fairly new but already impressively large body of research into this phenomenon, and the vast and terrible historical evidence of its workings and effects, in her book *The Self Beyond Itself*. She concludes:

> On the societal scale, our freedom lies in developing institutions and cultural beliefs and practices and families that shape our brains toward the greatest good rather than toward narrow interests, and toward health rather than addictive habits and other limitations, starting early in life.[18]

THE MYTH OF CREATIVE COMPETITION

In the Northern world, there has been a dominant idea that human nature is fundamentally competitive and individualistic.

Innovation is said to be driven by the lure of wealth; hence, if we want nice things like iPhones, we need an unequal society, where there is a chance to get ahead. But when we actually see innovation in action, that is not how it works.

Some of the clearest refutations of the 'spur of competition and profit' argument come from the world of computers, with its egalitarian, collaborative origins and continuing culture. This has even inspired a wave of wishful thinking, to the effect that computers herald a new, egalitarian age. The social-science writer David Berreby has described computer programmers as 'The hunter-gatherers of the knowledge economy'[19] and identifies a long list of similarities between the new knowledge-workers' behavior and value systems, and those of the hunter-gatherers described by anthropologists such as Christopher Boehm and Marshall Sahlins. 'Can we win the joys of the hunter-gatherer life for everyone?' he asks, 'Or will we replicate the social arrangements of ancient Athens or medieval Europe, where freedom for some was supported by the worst kind of unfreedom for others?'

Technology's history makes more sense if we recognize it as a constant, global, human activity, unconcerned with corporate or national boundaries, or the status systems within them. But as technologies became more powerful, elites became increasingly aware of them as threats or opportunities, and either suppressed them, or appropriated them and tried to channel their development in directions they found acceptable.

This fits better with innovators' own experience. One hardly ever hears of an important innovation emerging from a boardroom or a chief executive's office. Usually, the innovation emerges from an organization's nether regions, or from outside any recognized organization. The innovator must laboriously build up evidence, gather allies, pay court to financiers and backers, and only then, on a good day with a following wind, perhaps attract the boardroom's attention. Then, perhaps, the organization will adopt the innovation and perhaps, after modifications and compromises of various kinds, sell it to the world as yet another great product from Apple, Canon, or whoever.

More often than not the innovation is used, but without much appreciation. When the first, small capitalist states arose in 16th-century Europe, major innovations had quietly been emerging from within European towns, or making their way into Europe in informal ways, from China and India, for several centuries. The merchant elite did not acknowledge them officially until 1474, when the state of Venice started granting its first 10-year patents. To those who only look at the official record, this has suggested the start of a period of innovation, but 1474 more likely marked the beginning of the end of Europe's great period of innovation – mostly achieved by anonymous, federated craftworkers. In a major study of medieval industries published in 1991, Steven Epstein wrote:

> More than five centuries of increasingly effective patents and copyrights have obscured the medieval craft world in which such rights did not exist, where, to the contrary, people were obliged to open up their shops to guild inspection and where theft of technology was part of the ordinary practice of business.[20]

This allowed a capitalist myth to flourish, that there was no progress at all in either technology or in science in Europe from the end of the Roman Empire until the Renaissance. Lynn Townsend White, who became fascinated by this 'non-subject' in the early 1930s, wrote in 1978: 'As an undergraduate 50 years ago, I learned two firm facts about medieval science: (1) there wasn't any, and (2) Roger Bacon was persecuted by the church for working at it.'[21]

But between the 10th and 15th centuries, the stirrup, clockwork, glassmaking, the windmill, the compass, gunpowder, ocean-going ships, papermaking, printing and a myriad other powerful technologies were introduced or invented and developed under the noses of European elites, and were adopted and used by them greedily, ruthlessly and generally without comprehension. Many modern technologists and technology workers would say that little has changed.

Despite the contradictions, modern society is permeated by a belief that capitalism is pre-eminent when it comes to creating new technologies, and that computers and electronics have proved this beyond doubt. Even people on the Left say so. The sometime-socialist economist Nigel Harris has written of 'the great technical triumphs of capitalism – from the steam engine and electricity to the worldwide web, air travel and astronauts'.[22] He laments the environmental damage that seems to come with them, but he concedes that 'markets and competing capital have a spectacular ability to increase output and generate innovations'.

An eminent Marxist, the geographer David Harvey, says: 'The performance of capitalism over the last 200 years has been nothing short of astonishingly creative.'[23] A moderately left-of-center commentator, Jonathan Freedland, argues that, even though capitalism has led to the climate crisis,

> we would be fools to banish global business from the great climate battle... Perhaps capitalism's greatest contribution will come from the thing it does best: innovation.[24]

The idea is even, apparently, central to the theories of Karl Marx and Frederick Engels. Their *Communist Manifesto* of 1848 contains what a highly respected Marxist scholar, Michael Burawoy, calls 'a panegyric to capitalism's power to accumulate productive forces'. The Manifesto says:

> Subjection of nature's forces to man, machinery, application of chemistry to industry and agriculture, steam navigation, railways, electric telegraphs, clearing of whole continents for cultivation, canalization of rivers, whole populations conjured out of the ground – what earlier century had even a presentiment that such productive forces slumbered in the lap of social labor?

But are Marx and Engels telling us that capitalism is a Good Thing? Of course not. They hated capitalism and expressed their hatred for it with vigor, relish and creativity. Marx continually alluded to its vampiric qualities (inspiring Mark Neocleous to call capitalism 'the political economy of the dead'[25]). Marx often depicts capitalists as almost comical victims of circumstances.

Capitalism, for Marx, is something like a natural phenomenon that hubristic entrepreneurs unleash but can barely control, still less understand. Marxism's own parallel success story since 1848 surely stems to some extent from the way its explanation of grandiose capitalist behavior has rung so true, capturing the experience of so many millions of workers in so many different working situations.

But whatever Marxists think, conventional wisdom nowadays has it that capitalists are very wise, and that market competition between firms spurs innovation.

WHAT CAPITALISM CANNOT DO

A reputation for innovation started to become a valuable corporate asset around the time of the Second World War, and it has become almost an article of faith since then that modern, profit-driven capitalist firms, with their teams of highly motivated researchers, are the supreme exponents of technological innovation. Nonetheless, governments have occasionally felt the need to find out whether this really is the case or not.

A 1965 US Senate committee invited a succession of the leading authorities from all areas of industry to give them the benefit of their research into innovation, in an effort to decide whether the government should channel more of its research funding to large firms rather than small ones, and encourage business to concentrate into larger units, to foster a greater rate of innovation.[26]

The economist John Kenneth Galbraith, by no means an uncritical supporter of unfettered capitalism, had written not long before that 'A benign providence... has made the modern industry of a few large firms an almost perfect instrument for inducing technical change'. Other eminent experts, such as the education theorist Donald Schön, disagreed, citing a major study called *The Sources of Invention* by a British research group headed by the Oxford economist, John Jewkes.[27] This had seriously challenged the credibility of the corporate approach to major scientific challenges, with its emphasis on teamwork and targets – an approach equally prevalent both in the USSR and in the capitalist countries. Jewkes examined industries such as radar, television, the jet engine, antibiotics, human-made fibers, steel production, petroleum, silicones and detergents. The USSR came out badly from Jewkes's study (no important innovations in any of the areas examined) but then, so did capitalist firms. In every area studied, innovation had dried up from the moment capitalist firms took a serious interest in it.

The Senate committee asked one of Jewkes' co-authors, David Sawers, for an update on his study of the US and European aircraft industries. Sawers had found lots of growth, but not much serious innovation. US aviation had not come up with anything very new since the Second World War and it was still living off a few, mainly German, inventions that had been made in wartime. Almost none of the major advances in aircraft design, anywhere in the world, had come from private firms, and firms had been particularly resistant to jet propulsion. Jet airliners, he said, had only become established thanks to the US government underwriting the development costs and guaranteeing a military market for the Boeing 707. Major advances such as streamlining, swept-back wings, delta wings, and variable geometry all came from outside capitalist firms and had had a job being accepted by them – unless underwritten by military contracts. The only significant pre-War improvements made by capitalist firms that he had been able to find were the split flap (invented by Orville Wright, an old-school inventor, so not exactly representative) and the slotted flap (introduced by Handley Page in the UK). After the War, some modest innovation had been done by European aircraft firms on delta wings – the least adventurous of the new geometries (this work eventually led to the Concorde supersonic airliner, which was built largely at public expense, as a prestige project).

In the steel and automobile industries it was the same story: in general, no innovation except with lots of government support, or via the dogged persistence of independent inventors. In the photographic industry, Kodachrome (the first mass-market color film, launched in 1935) was literally invented in the kitchen sink by two musicians, Leopold Godowsky Jr and Leopold Mannes, in their spare time. The two men had struggled at the project largely at their own risk since 1917.

Sawers' colleague, Richard Stillerman, put it thus:

> Making profits is the primary goal of every firm. Few, if any, firms would support the kind of speculative research in manned flight undertaken by the Wright Brothers at the turn of the century after experts proclaimed that powered flight was impossible. Or the risky experiments on helicopters which a horde of optimistic individuals carried forward over several decades. Or the early rocket research pursued by individuals with limited financial backing.[28]

Turning to electronics, the committee learned that one of the industry's greatest success stories, xerography (the technology behind the huge Xerox Corporation), had only seen the light of day after its inventor, Chester Carlson, approached the non-profit Battelle Memorial Institute for support. Other major innovations had been actively resisted by the firms in which they were being developed. Arthur K Watson (son of Thomas J Watson, founder of the IBM Corporation) was quoted to the effect that:

> The disk memory unit, the heart of today's random access computer... was developed in one of our laboratories as a bootleg project – over the stern warnings from management that the project had to be dropped because of budget difficulties. A handful of men ignored the warning ... They risked their jobs to work on a project they believed in.[29]

The committee learned that talented researchers were fleeing capitalist firms to set up or join small, independently funded outfits where they could develop their ideas without interference. The 'small startup' subsequently became one of the iconic conventions of the electronics/computer industry and was touted as a great success, but small startups didn't, don't and can't carry out the sustained research effort publicly funded teams are capable of. While some get rich, the vast majority do not.[30]

INNOVATION IN THE 'NEW ECONOMY'

In the early 1980s the British government faced concerns about the country's ability to compete in what was tentatively being called 'the new economy'. Not all experts shared the government's deep conviction that more intense commercial competition would deliver the requisite innovations, so they commissioned two US academics to nail the matter: Nathan Rosenberg and David Mowery.

Mowery examined nine major pieces of research into industrial innovations that claimed to have been inspired by market demand. On close inspection, he found that, while it was true that they had arisen within firms, most of them had been the fruit of researchers following their own interests, and 'the most radical or fundamental ones were those least responsive to "needs"'.[31] He also explained that most of the key innovations in computing and electronics had happened well outside the reach of the market: in universities and government research organizations. He concluded that 'while one may rely upon the ordinary forces of the marketplace to bring about a rapid diffusion of an existing innovation with good profit prospects, one can hardly rely completely upon such forces for the initial generation of such innovations'.

Another big study, published in 2004 by Daniel Cohen and colleagues for the Rodolfo Debenedetti Foundation,[32] also found 'scant' evidence of any link between competition and innovation,[33] although any innovations that were adopted did seem to be diffused more rapidly by competition – the same point that Mowery had made 20 years earlier.

The study also found that the innovations carried out by individual firms tended to consist of adding 'tweaks' of their own, preferably patentable ones, designed to secure a positional edge in the market and perhaps monopolize some particular aspect of it. The story of the World Wide Web (now known simply as 'the Web') is a classic demonstration of how difficult it then becomes to preserve the innovator's original vision, and the surprisingly large environmental cost of departing from it – this is described in detail in Chapter 10.

WHY CAPITALISM INHIBITS INNOVATION

The 19th-century artist, writer and socialist William Morris described capitalist competition as 'a mad bull chasing you over your own garden'[34] – a view now supported by a wealth of research. Governments and firms started to take interest in creativity as the pace of technological change accelerated after the Second World War. Creativity became a major area of study, which has consistently found that even mildly competitive environments and situations of unequal power are completely incompatible with creative thinking.[35]

In one of many experiments described in a popular book about creativity by Guy Claxton, two groups of non-golfers were taught putting.[36] Both were instructed in exactly the same way, but one group was told that they'd be inspected at the end of the course by a famous professional. This group's performance (measured in balls successfully 'sunk') was far lower than the other group's. The small anxiety of knowing a professional would be watching them devastated their ability to learn.

Firms became very keen on using the new research to teach their staff to be more creative, and some psychologists found creative ways to teach creativity, without challenging the creativity-stifling structures of the firms that paid them. Brainstorming, team away-days and the creative enclaves mentioned earlier, are all products of this 'creativity movement'.

Claxton has an illuminating anecdote about George Prince, co-founder of a popular 'creativity-enhancing' system called Synectics. To Prince's chagrin, his own research led him to realize that the business context made his enterprise hopeless:

> Speculation, the process of expressing and exploring tentative ideas in public, made people, especially in the work setting, intensely vulnerable, and that… people came to experience their workplace meetings as unsafe.
>
> People's willingness to engage in delicate explorations on the edge of their thinking could be easily suppressed by an atmosphere of even minimal competition and judgement. 'Seemingly acceptable

> actions such as close questioning of the offerer of an idea, or ignoring the idea ... tend to reduce not only his speculation but that of others in the group.'[37]

Even positive motivation stupefies – in particular, financial reward. In 1984 James Moran and his colleagues not only showed that reward impaired performance, they also discovered *how* it impaired performance: essentially by causing 'a primitivization of psychological functioning'. The subjects regressed in effect to childhood, and performed below their mental age.[38]

After the 2007/8 financial crisis, the Nobel economics laureate and psychologist Daniel Kahneman wrote a global bestseller, *Thinking, Fast and Slow*, which explained exactly why performance bonuses do not and cannot work – and why the decisions of even the brightest corporate leaders are generally governed more by luck and delusion than by genius. [39]

Hierarchy invokes a different mindset from the one we enjoy among equals, and in which we are most productive and creative. In his 2009 book, *The Master and His Emissary: the Divided Brain and the Making of the Western World*, Iain McGilchrist has shown how this ties in with the human brain's well-known (but subtle) 'lateralization'. The brain's two halves have somewhat different functions but the most important difference is one of style. In simple terms: 'The right hemisphere has an affinity for whatever is living, but the left hemisphere has an equal affinity for what is mechanical'.[40] The two halves need to work in tandem but the peculiar dynamics of Western society create over-reliance on the left hemisphere, with its more intense, focused approach, resulting in:

> a pathological inability to respond flexibly to changing situations. For example, having found an approach that works for one problem, subjects [who have suffered damage to the right side of the brain] seem to get stuck, and will inappropriately apply it to a second problem that requires a different approach – or even, having answered one question right, will give the same answer to the next and the next.[41]

In his concluding chapter, McGilchrist imagines what the world would look like to a brain consisting of nothing but its left hemisphere, and it looks a lot like the world where so many of us have to live and work: it recognizes skill only in terms of what can be codified, is obsessed with details at the expense of the broader picture, has little empathy. In this not-entirely-hypothetical world:

> Fewer people would find themselves doing work involving contact with anything in the real, 'lived' world. Technology would flourish, as an expression of the left hemisphere's desire to manipulate and control the world for its own pleasure, but it would be accompanied by a vast expansion of bureaucracy, systems of abstraction and control. The essential elements of bureaucracy, as described by Peter Berger and his colleagues [in the book The Homeless Mind, 1974], show that they would thrive in a world dominated by the left hemisphere.[42]

Capitalist industry is aware something is amiss, but cannot contemplate the obvious solution (stop being capitalist) and it is not in the creativity consultant's interest to point it out either.

CAPITALISM DIDN'T MAKE COMPUTERS...

The notion that computers demonstrate capitalism's creativity is finally belied by the computer's history. Capitalism ignored computers for more than a century, and we might still be waiting for them now, had it not been for brief moments of egalitarian collaboration, grudgingly or accidentally tolerated by the elites, during the worst crises of the Second World War. And even then, capitalism had to be laboriously spoon-fed the idea for a further decade and more, before it would invest its own money in it. Mariana Mazzucato's book *The Entrepreneurial State* reveals the enormous scale of capitalist industry's dependence on publicly funded research. One of her case studies is Apple's iPhone, which would amount to nothing very much without the billions of dollars' worth of research effort that produced everything from its fundamentals (the microprocessors, computer science itself) to its most modern-looking features: the touch screen, the Global Positioning System (GPS), its voice-activated SIRI 'digital assistant', and the internet itself: all the 'features that make the iPhone a smartphone rather than a stupid phone'.[43]

Programmable, digital computers in the modern sense (but using mechanical gears rather than electronic valves) had been possible since the 1830s, when the British mathematicians Charles Babbage and Ada Lovelace were developing the science behind them. Babbage even developed much of the necessary hardware which, when partially completed (using 1830s tools and methods) in the 1980s, worked perfectly well. By most logical considerations, there was abundant need for Babbage's machines in Britain's burgeoning commercial empire – but human labor was dirt cheap and British commerce simply wasn't interested in them.[44]

Computer scientist and historian Brian Randell, founder of the IEEE's *Annals of the History of Computing*, found other perfectly viable ideas for computers that should have been snapped up by a capitalist system that actually did what it claimed to do, namely foster innovation and drive technological progress. One of his discoveries was Dublin accountant Percy Ludgate's 1907 proposal for a portable computer that could have multiplied two 20-digit numbers in less than ten seconds, using electric motor power. Its specification included the ability to set up sub-routines. Ludgate's work attracted great interest in the learned societies in Dublin, London and internationally, and the British Army eventually hired him to help plan logistics during the First World War, but he drew a complete blank with the businesses that would have benefited most from his invention.[45]

The Smithsonian Institution's computer historian, Henry Tropp, has written:

> We had the technical capability to build relay, electromechanical, and even electronic calculating devices long before they came into being. I think one can conjecture when looking through Babbage's papers, or even at the Jacquard loom, that we had the technical ability to do calculations with some motive power like steam. The

realization of this capability was not dependent on technology as much as it was on the existing pressures (or lack of them), and an environment in which these needs could be sympathetically brought to some level of realization.[46]

It took the exceptional circumstances of a second global war, topped by the threat of a nuclear one, to nudge governments and managements in the advanced nations into providing or at least tolerating briefly the kinds of environments where 'these needs could be sympathetically brought to some level of realization': highly informal settings where machines like the Colossus were built (by Post Office engineers for the British code-breaking center at Bletchley Park, to break German high-command codes – but scrapped and erased from the official record immediately afterwards[47]), and the 'Electronic Numerical Integrator And Computer' (ENIAC), designed in Philadelphia for calculating gunnery tables, and completed in 1946.

Colossus was the world's first true, programmable, digital electronic computer, and it owed its existence to suspension of 'business as usual' by the threat of military defeat, which also briefly overshadowed the normal regime of homophobia and snobbery. Bletchley Park's codebreaking genius, the mathematician Alan Turing, was tolerated while hostilities lasted despite his homosexuality and awkwardness. Colossus was built by a team of five working-class General Post Office (GPO) engineers who would never have been allowed near such an important project in normal times (they very nearly weren't anyway, and certainly weren't as soon as the War ended).

The GPO team was led by TH (Tommy) Flowers, a bricklayer's son from the east end of London. While in his teens, Flowers had earned an electrical engineering degree by night while serving a tough engineering apprenticeship during the day. By 1942 he was one of the few people in the world with a practical and theoretical knowledge of electronics, and an imaginative grasp of its possibilities. Management referred to Flowers as 'the clever cockney', and tried to get him off the project, but he and Turing got on well from the first, and Turing made sure the project went ahead despite the opposition and sneers. One of Flowers' team, SW Broadhurst (a radar expert who had originally joined the GPO as a laborer) took it all without complaint. Computer scientist Brian Randell recorded this impression from him in 1975:

> The basic picture – a few mathematicians of high repute in their own field accidentally encounter a group of telephone engineers, of all people... and they found the one really enthusiastic expert in the form of Flowers, who had a good team with him, and made these jobs possible, with I think a lot of mutual respect on both sides. And the Post Office was able to supply the men, the material and the maintenance, without any trouble, which is a great tribute to the men and the organization.[48]

Turing's biographer and fellow-mathematician, Andrew Hodges, records that Colossus was built with extraordinary speed and worked almost perfectly, first time: 'an astonishing fact for those trained in the conventional wisdom. But in 1943 it was possible both to think and do the impossible before breakfast.'[49] The GPO team worked so fast that much of the first Colossus ended up being paid for, not by the government, but by Flowers himself, out of his own salary.[50]

But as soon as the War was over these men were sent back to their regular work and could make no further contribution to computing, or even talk about what they had done, until the secret finally emerged in the 1970s. As Flowers later told Randell: 'It was a great time in my life – it spoilt me for when I came back to mundane things with ordinary people.'

...BUT TOOK COMPUTING DOWN THE WRONG PATH

The ENIAC was nearly a dead end for almost the opposite reason. Its designers, Presper Eckert and John Mauchly, were so intent on being successful capitalists that they nearly buried the project themselves. Unlike Colossus, ENIAC gained public recognition but, as the Dutch computer scientist and historian Maarten van Emden has argued, the rapid commercialization that its makers had in mind could have turned it into a 'revolution that didn't happen'[51] had it not been for the fortuitous and somewhat unwelcome involvement of the Hungarian mathematician John von Neumann who (by further fortuitous connections, including discussions years earlier with Alan Turing in Cambridge) was able to relate what he saw to other, apparently quite unrelated and abstruse areas of mathematics and logic.

Eckert and Mauchly seem not to have understood von Neumann's idea which, nonetheless, von Neumann was able to publish, to their annoyance, free of patent restrictions, in the widely circulated report on ENIAC's successor, the EDVAC. This report effectively kept computer development alive, and out of the hands of normal capitalist enterprise, which (as Van Emden argues) would then have smothered it. He writes that:

> *Without von Neumann's intervention, Eckert and Mauchly could have continued in their intuitive ad-hoc fashion to quickly make EDVAC a success. They would also have entangled the first stored-program computer in a thicket of patents, one for each ad hoc solution. Computing would have taken off slowly while competitors chipped away at the initial monopoly of the Eckert-Mauchly computer company. We would not have experienced the explosive development made possible by the early emergence of a design that, because of its simplicity and abstractness, thrived under upheaval after upheaval in electronics.*

Von Neumann's idea – the 'von Neumann architecture' – specified a central unit for doing arithmetic; a memory store shared by the program instructions, the data to be worked on and intermediate results; and a control unit to initiate each step of the program, copying data and instructions alternately from and back to memory in a 'fetch-execute cycle', as well as receiving input from a keyboard (or other input) and passing it back to a printer (or other output). The system is robust and comprehensible because it does just one thing at a time, in step with a timing pulse or 'clock'. This turned out to have surprisingly expensive consequences, as computers began to be applied to tasks never envisaged in the 1940s: taking photographs and movies, playing music, and so on... as we shall see later on, in Chapter 11.

Von Neumann's design is still the basis of nearly all modern computers, and the whole computer revolution might not have happened, had it not been for his freakishly broad interests, his unwanted intervention in Eckert and Mauchly's business, and then his airy disregard of commercial propriety in circulating his specification. This proved a lucky break for capitalism, despite itself.

More and more powerful machines became possible thanks to von Neumann's disruptive presence, but capitalist firms still resolutely had nothing to do with their development unless all of the costs were underwritten by governments. As for using computers themselves, they had to be coaxed endlessly, like recalcitrant children, before they would even try what was good for them. Computer development remained utterly dependent on government support for decades.

In his 1987 study for the Brookings Institute, Kenneth Flamm estimated that in 1950 more than 75 per cent of US computer development funding had come from the government, and any commercial investments were largely made in anticipation of lucrative defense contracts. One such contract financed development of IBM's 701 machine – originally known as 'The Defense Calculator'. IBM's commitment to computers was built on guaranteed returns from military projects. A decade later, in 1961, the US government was still funding twice as much computer research as the private sector did, and this remained largely the pattern through the Cold War era. Another historian, Paul Edwards, has wondered whether digital computers would even have survived had it not been for the Cold War.[52]

Personal computers (from which iPhones and their like are descended) might have remained a quaint, hobbyist idea had today's commercial norms been in place in the 1970s. The idea was shaped to a great extent by political activists opposed to big business and the military,[53] and the world of business scorned them until the first computer spreadsheet (the 'magic piece of paper' that recalculates your sums for you, when you change any of the figures) appeared in 1979. This was Dan Bricklin's Visicalc, which he wrote for the Apple II computer, giving it a desperately needed foothold in the business market. Bricklin did not patent the spreadsheet idea – and could not have done until two years later, by which time the idea had been picked up by other software companies. The excitement about spreadsheets contributed to IBM's hurried but decisive decision to enter the personal computer market in 1981.

When firms finally discovered the computer's benefits, there was a competitive frenzy. As I will argue in later chapters, this led to a wholesale restructuring and concentration of the economy that somehow yielded very little beneficial effect on standards of living, but a great increase in human inequalities and impacts.

Many firms fell by the wayside along with whole areas of employment. So, too, did whole areas of technical possibility. As firms started to make money from selling computers, competitive development proceeded at such breakneck speed that attempts to open up new architectural possibilities were bypassed before they could be made ready for general use. We will look at some of these in later chapters.

Market forces effected a swift and radical simplification of what people thought of as 'the computer', forcing its development to be channelled along a single, very narrow path. As we will see, von Neumann architecture became the 'only show in town' – a development that would have made von Neumann himself despair, and which has had surprising environmental consequences. The rich diversity of technologies that characterized computing before that time evaporated, leaving a single, extremely inefficient technology to serve everything from mobile phones to audio equipment to financial markets. The computer revolution, which promised to enrich human lives and reduce human impacts, and could certainly have done so, in the event did the exact opposite.

1 'Another suicide at Foxconn after boss attempts damage control', *China Labour Bulletin,* 27 May 2010, nin.tl/FoxconnCLB (retrieved 01/06/2010).
2 Nick Cohen, 'How much do you really want an iPad?' *The Observer*, 30 May 2010.

3   Milanovic, Williamson and Lindert, 'Measuring Ancient Inequality', *NBER working paper*, October 2007.
4   George Packer, 'Change the World: Silicon Valley transfers its slogans – and its money – to the realm of politics', *The New Yorker*, 27 May 2013.
5   Glynn Moody, *Rebel Code: Linux And The Open Source Revolution*. Basic Books, 2009, p 28.
6   GNU is a 'recursive acronym' for 'GNU's not Unix'. Unix, although a proprietary product, was also largely the work of another lone programmer, Ken Thompson, who wrote it in his own time, against the wishes of management, in 1969.
7   Moody, op cit.
8   W Brian Arthur, *The Nature of Technology: What It Is and How It Evolves*, Free Press, 2009.
9   J Gleick, *The Information: A History, a Theory, a Flood*, HarperCollins Publishers, 2011.
10 Arthur, op cit., p 25.
11 Richard Rudgley, *Lost Civilisations of the Stone Age*, Century, London, 1998.
12 Martin Jones, *Feast: Why Humans Share Food*, Oxford University Press, 2007.
13 Hugh Brody, *The Other Side of Eden,* Faber, London, 2001.
14 George Dyson, *Baidarka*Alaska Northwest Pub Co, 1986.
15 Christopher Boehm, *Hierarchy in the Forest: the evolution of egalitarian behavior*, Harvard University Press, 2001.
16 M Sidanius & F Pratto, *Social Dominance: an intergroup theory of social hierarchy and oppression*, Cambridge University Press, 1999.
17 George Packer, 'Change the World', *The New Yorker*, 27 May 2013, nin.tl/packerNY.
18 Heidi M Ravven, *The Self Beyond Itself,* New Press, 2013, p 299.
19 David Berreby 'The Hunter-Gatherers of the Knowledge Economy,' *Strategy and Business* 16 July 1999.
20 Steven A Epstein, *Wage Labor and Guilds in Medieval Europe*, University of North Carolina Press, 1991, p 245.
21 Lynn Townsend White, *Medieval religion and technology: collected essays*, Publications of the Center for Medieval and Renaissance Studies, University of California Press, 1978, pp x-xi.
22 Nigel Harris, 'Globalisation Is Good for You', *Red Pepper*, 3 Dec 2007, nin.tl/HarrisRP
23 David Harvey, *The Enigma of Capital,* Profile, London, 2010, p 46.
24 Jonathan Freedland, *The Guardian*, 5 Dec 2007, nin.tl/Freedland07
25 Mark Neocleous, 'The Political Economy of the Dead: Marx's Vampires', *History of Political Thought*, Vol 24, No 4, 2003, pp 668-84.
26 *Economic Concentration,* 'Hearings before the subcommittee on antitrust and monopoly of the committee of the judiciary', US Senate, 89th Congress, First Session, 18, 24, 25 and 27 May and 17 June 1965.
27 John Jewkes, *The Sources of Invention*. Macmillan, London/ St Martin's Press, NY, 1958.
28 *Economic Concentration*, op cit, p 1075.
29 Ibid, p 1217.

30 'Silicon Valley's culture of failure ... and 'the walking dead' it leaves behind', Rory Carroll, *The Guardian*, 28 June 2014, nin.tl/Siliconfailure
31 N Rosenberg, *Inside the Black Box: Technology and Economics*, Cambridge University Press, 1983, p 229.
32 Daniel Cohen, P Garibaldi and S Scarpetta. *The ICT Revolution: Productivity Differences and the Digital Divide*, Oxford University Press US, 2004.
33 Ibid, p 85.
34 William Morris, *How we live and how we might live*, 1884, 1887.
35 *Creativity. Selected readings*, ed Philip E Vernon, Penguin, 1970.
36 Guy Claxton, *Hare Brain, Tortoise Mind*, Fourth Estate, London, 1998.
37 Ibid, pp 77-78.
38 James D Moran, 'The detrimental effects of reward on performance', in Mark R Lepper and David Greene, *The Hidden Costs of Reward*, Erlbaum 1978; James D Moran and Ellen YY Liou, 'Effects of reward on creativity in college students of two levels of ability', *Perceptual and Motor Skills*, 54, no 1, 1982, 43–48.
39 Daniel Kahneman, *Thinking, Fast and Slow*, Farrar, Straus and Giroux, 2011.
40 Iain McGilchrist, *The Master and His Emissary*, Yale University Press, 2009, p 55.
41 Ibid, pp 40-41.
42 Ibid, p 429.
43 Mariana Mazzucato, *The Entrepreneurial State: debunking public vs. private sector myths, 2014*. Mazzucato published an earlier, shorter version for Demos in 2011, which may be downloaded free from their website: nin.tl/Mazzucato
44 J Gleick, *The Information: A History, a Theory, a Flood*, HarperCollins, 2011.
45 Brian Randell, 'Ludgate's analytical machine of 1909', *The Computer Journal*, vol 14 (3) 1971, pp 317-326.
46 Henry Tropp, quoted in B Winston, *Media Technology and Society, a history: from the telegraph to the internet*, Routledge, London, 1998.
47 Andrew Hodges, *Alan Turing: The Enigma*, Simon and Schuster, 1983; G Dyson, *Turing's Cathedral*, Knopf Doubleday, 2012; Gordon Welchman, *The Hut Six Story*, McGraw-Hill, 1982.
48 Brian Randell, 'The Colossus', in *A History of Computing in the Twentieth Century*, ed N Metropolis, J Howlett and G-C Rota, Academic Press, 1980.
49 Hodges, op cit, p 268.
50 Randell, 'The Colossus', op cit.
51 Maarten van Emden, 'The H-Bomb and the Computer, Part II', *A Programmers Place*, accessed 19 Aug 2014, nin.tl/EmdenHBomb
52 PN Edwards, *The Closed World*, MIT Press, 1996.
53 John Markoff, *What the Dormouse Said: How the Sixties Counterculture Shaped the Personal Computer Industry*, Viking, 2005.

2
From water mills to iPhones: why technology and inequality do not mix

The middle of the 20th century saw the world move in the direction of greater equality, including the rise of the welfare state – and that produced egalitarian hopes for computing. But these were dashed by the neoliberal policies from the 1970s onwards, which saw a return to medieval levels of social inequality. Medieval history shows what happens when new technologies are introduced into very unequal societies, as with the water mill and then the gun. The precedents are not encouraging.

Computers were one of a constellation of transforming developments that broke through into the real world during the politically contested and chaotic years during and just after the Second World War, and not just in the 'hard sciences'.

Old shibboleths about the nature and potential of human beings, how they should be treated, and about how society should be ordered, were being demolished and discredited by new currents of work in the social sciences (anthropology, psychology, neuroscience, education) and by people's own direct experiences in the War. Understandings of how the planet itself worked were being elaborated at a breakneck rate with the brand-new sciences of cybernetics and general systems theory. Many people thought that these understandings, which had developed in lockstep with the development of computing machinery and signal processing, were much more significant for humanity's future than the machinery itself. The world was being recognized as a whole system, and so was knowledge.

All economic and political trends seemed to be pointing toward greater and greater equality. The brief failure of 'business as usual' that allowed the likes of Alan Turing and Tommy Flowers to work together was one of millions of instances of a much bigger egalitarian and humanistic turn that was struggling to shape the course of events globally.

Old and entrenched doctrines of innate inequality and racial superiority, so widely held and even respectable before the war, were abruptly silenced by the defeat of the Fascist powers, and the exposure of the unambiguous horror of the Nazi 'final solution' to its 'Jewish problem'. Fascism ceased to be respectable, and there was a general stampede among those who had supported it during its heyday, to higher moral ground. The concept of 'crimes against humanity' gained legal force for the first time, and political leaders were jailed, put on trial and sentenced for committing them.

The post-War political climate was shaped to a surprising degree by people who had wielded pathetically little military muscle during the conflict: resisters and rescuers now found themselves almost the sole uncompromised bearers of the West's claims to moral superiority. Leftwingers and liberals, who had been pre-eminent in the resistance movements, briefly found themselves at or at least in sight of the top table. One of these was the late Stéphane Hessel, author of the 2011 bestseller *Time for Outrage!*[1] (published in French as *Indignez-vous!*), which sold millions of copies worldwide, especially among supporters of the various mass protest movements that broke out after the financial crash of 2007/8.[2] Hessel was a German-born, Jewish member of the French resistance who survived capture and torture, escaped, and ended the war as part of the team responsible for drafting the Universal Declaration of Human Rights of 1948.

While France was still under Nazi occupation, its National Resistance Council (the CNR) drafted a Charter demanding, for example,

> *a complete Social Security scheme aimed at guaranteeing means of subsistence for all citizens, in every case where they are unable*

> to procure these through labor, the management of which will belong to representatives of the interested parties and the state [and] a pension that will allow old workers to end their days with dignity.

The Resistance was rudely sidelined by the incoming De Gaulle government, but even so, the extent to which its proposals were implemented is striking (especially when compared with social provisions at the beginning of the 21st century).

The Scottish writer Neal Ascherson has observed that, independently of each other, in 'the years between, say, 1943 and 1947... resistance movements from Poland to France and southern Europe developed similar programs for revolutionary democratic renewal after the war'.[3] Britain's welfare state was part of the phenomenon, perhaps reflecting a feeling many British people had of having in a sense lived 'under occupation' by Winston Churchill – whom they comprehensively ejected from office when elections were resumed in 1945.

All of the victorious powers introduced some kind of basic welfare provision for their citizens: pensions, unemployment benefits, healthcare and even that essential preserve of the rentier class, housing.

Major firms and industries were nationalized with popular approval. On France's liberation, the owner of the Renault automobile company, Louis Renault, was arrested as a collaborator and his company was confiscated. The following year the rest of the automobile industry, the whole banking sector and the coalmining industry were nationalized.

The Right was even more deeply and durably discredited in Germany, where strong social-justice measures were introduced after 1945. Trade unions were given a stronger role, including seats in German boardrooms. The French economist Thomas Piketty has argued that this has helped maintain the country's prosperity by restraining the scope for runaway managerial incomes.

The social groundwork was set for a progressive shift towards greater and greater equality – in line with the uncontested general acceptance that all human beings were, in fact, equal.

The world was suddenly full of people who had discovered capacities for heroic and creative achievement, and interests and experiences they never would have dreamed possible a few years earlier. Women in their tens of thousands had escaped from domestic duties, done 'men's jobs', including some of the most dangerous and exciting ones. There were thousands of people like Tommy Flowers who had lived and worked at the outer edge of technological and human possibility, and resented being dragged back into their allotted social niches and told to stay there.

The conditions needed for people to realize their potential became, however, a legitimate subject of research. Education was expanded massively in all the main economies, and there was a proliferation of new, more 'liberal' educational regimes. For the first time in centuries, it was becoming official policy that all human beings should be treated humanely, and this was backed by new international laws. There was also a widespread sense that the future would be one of greater equality within and between nations, rather than less.

It is true that much of the new work in psychology served a social-control agenda – such as harnessing propaganda and finding better ways of getting troops to follow orders – but it also included radical departures in, for example, understanding child development and mental illness.

Both during and after the War governments and aid agencies were overwhelmed with traumatized people and orphans. In Britain, Wilfred Bion pioneered what became the group therapy movement for the treatment of mental distress, when the War Office found itself in charge of thousands of men who had been emotionally shattered by the ordeal of Dunkirk.[4] While working with evacuee and refugee children, John Bowlby developed his theory of attachment and the damage caused by separation, which profoundly altered views about parenting and the treatment of children in the post-War years. Bowlby's colleague Donald Winnicott developed a theory of the vital importance of play.[5] These theories have subsequently done nothing but grow in their explanatory and therapeutic power.

Genocide, and the Nazi defeat, triggered a wave of research

into the human capacity for cruelty. Why do apparently decent people, in some circumstances, become murderers and torturers, while others feel compelled to become rescuers and resisters, whatever the risks?

Theodor Adorno's colleagues in New York discovered plenty of potential for complicity in brutal regimes among ordinary Americans.[6] Through scores of interviews, they identified a strong link between strict, punitive parenting, and attitudes predisposing people to support fascistic regimes and authoritarianism in general. Stanley Milgram[7] and others began to expose the unwelcome truths about the ease with which people can be persuaded to inflict lethal harm on others.[8]

THE 'GREAT COMPRESSION'

This was, above all, the period in which income and wealth differentials began to shrink; it has been called 'the great compression'.[9] It is still not entirely clear when or how the compression began, but it certainly ended in the 1970s, when the ascendance of neoliberal economic policies threw the process into reverse, initiating what subsequent economists have called 'the great divergence' in incomes and wealth.[10]

Progressive taxation was adopted throughout the West – at far higher levels than had ever been contemplated before, and not always by left-wing governments. A proposal for a mere two-per-cent extra tax on the wealthiest had been howled down and dismissed out of hand in the France of 1914, but a rate of 50 per cent was successfully introduced as early as 1920 – by a reactionary rightwing government that sensed trouble otherwise.[11] The US led the rise in top tax rates during the war until the 1970s. In 1972, the Democrat presidential candidate George McGovern even proposed raising the top rate from 77 per cent to 100 per cent, to support a guaranteed minimum income. He was defeated, however, by the incumbent, Richard Nixon.[12]

The Second World War obviously necessitated a great levelling simply so that warring nations could overcome the problems of supply – for example, during the U-boat war in the Atlantic. But a levelling tendency had already been visible in France in 1920, and in the US after the financial crash of 1929. Demand for large, luxury cars fell abruptly; labor historian Rob Rooke records that demand fell steadily from 150,000 sold in 1929 to only 10,000 a year by 1937. Perhaps this was because it no longer felt right to drive in luxury when people were starving, and perhaps, following rumored or actual attacks on wealthy limousine-owners, rich people suddenly felt 'that ostentatious displays of wealth could cost them their lives'.[13]

This was part of a shift that prepared the moral climate for the New Deal reforms of Franklin D Roosevelt, who was elected by a landslide in 1932 while a socialist anthem, Yip Harburg's 'Brother, Can You Spare A Dime', stood for months at number one in the popular music charts.

The change had an enduring effect even on hardliners like General Douglas MacArthur, who earned public vilification in early 1932 and helped turn the tide of US opinion in favor of Roosevelt when he used tanks to crush a protest in Washington by the 'Bonus Army' of destitute World War One veterans and their families. In 1946, MacArthur took charge of Japan's occupation and reconstruction. This involved brutal suppression of leftwing forces but MacArthur also pushed through some surprisingly radical reforms, which have helped to maintain Japan's relatively greater equality today: land reform; a massive unionization campaign among industrial workers (albeit preceded by suppression of radical unions); the breaking-up of financial, industrial and media monopolies; and even subsidies for minority media.

The levelling trend was also galvanized by the sheer fact of a large, powerful and widely appealing rival value system in the USSR and, after the War, a greatly expanded Communist bloc of countries. Whatever the conditions of life actually were in those countries, their vigor and resilience were unquestionable, and, however corrupt their governing systems may or may not have been, they were propelled by a cause for which people were prepared to suffer and die, and which also inspired millions of people within the capitalist world. Capitalism's frontiers had become very close.

The Soviet ambassador to London, Ivan Maisky, described in his personal diary how, in cinemas after the Soviet victory at Stalingrad in 1943,

> [Stalin's] appearance on the screen always elicits loud cheers, much louder cheers than those given to Churchill or the king. Frank Owen [editor of the London Evening Standard] told me the other day... that Stalin is the soldiers' idol and hope. If a soldier is dissatisfied with something, if he has been offended by the top brass, or if he resents some order or other from above, his reaction tends to be colorful and telling. Raising a menacing hand, he exclaims: 'Just you wait till Uncle Joe gets here! We'll get even with you then!'[14]

Communist combatants did not always identify so readily with Stalin, especially those who had fought in the International Brigades for the Spanish Republic and seen Stalin's brutal and counter-productive takeover of the anti-fascist forces. Stalin's 1938 pact with Hitler had disastrously undermined French and German communists' struggles against fascism at a critical time.

After the War the French Resistance was sidelined and belittled by the brusque and professional-looking government of General Charles De Gaulle but it was hard to ignore the fact that the unquestioned heroes of the situation had not been the men in the impressive uniforms, but those much more ordinary people who had faced and suffered hardship, torture and death in the Resistance, or who had saved others from deportation and murder.

The name of Jean Moulin, the Resistance co-ordinator tortured to death by the Lyon Gestapo in 1943, was given to thousands of streets, squares, public parks and schools. None were dedicated to Pierre Laval, the Vichy prime minister who had assured the Nazi bureaucrats that 'our Jews are your Jews', and enforced targets for roundups and deportations that put many of his German opposite numbers to shame.[15] Yet thousands and perhaps millions of French people still regarded Laval as a responsible politician, and many of his colleagues continued to enjoy distinguished careers. Nonetheless, by the end of the Second World War, governments almost everywhere seemed to be embroiled in a conscious competition to occupy the 'moral high ground'.

The greater equality prevailing in the Scandinavian countries and in western Europe in the post-War period has been attributed to the sobering effect on those countries' elites of a well-armed communist country just over the border.[16] World Bank researchers deduced that a similar phenomenon was at work in the countries known in the 1990s as 'Asian Tigers' (Japan, South Korea, Taiwan, Singapore, Hong Kong, Thailand, Malaysia and Indonesia). In a 1993 report, *The East Asian Miracle*, the Bank's economists described how these countries had achieved big reductions in inequality between 1960 and 1980, and concluded that the policies had been in response to 'crises of legitimacy', arising from the presence nearby of communist rivals.[17]

EGALITARIAN HOPES FOR COMPUTING

In various places, but notably in Scandinavia, this moral shift not only affected social organization but also influenced global computer development.

In 1946 Norway's post-liberation government introduced a full-blooded regional policy not unlike the one adopted in 1959 by a very different-looking government in Cuba after the overthrow of the Batista regime: no community, no matter how small or remote, was to be denied the full range of public services or work opportunities enjoyed by towns; rural industries were to be protected. The computer pioneer Kristen Nygaard became national spokesperson for this policy in the 1990s:

> After the last world war we... decided that Norway should not deteriorate into a few densely populated areas around the large cities. Norway should exist as an interplay between vital local societies scattered all over the country. [... We] must keep our scattered settlement pattern, with an infrastructure covering and linking together all populated areas in the country. We cannot 'turn on and off' the local societies at will. A population returning to depopulated districts after a couple of decades will meet a

> deteriorated infrastructure, and will suffer from the loss of very important 'tacit knowledge' accumulated about the use of the land and resources during centuries of uninterrupted use and transfer of knowledge.[18]

For Nygaard, this was part of a radical politics that began in his childhood under Nazi occupation and which informed his approach to computerization. He and Ole-Johan Dahl invented the now-standard technique known as Object-Oriented Programming in 1965 specifically to support that kind of policy. Their computer language, SIMULA, was the starting point for most of the computer languages in use today, but it was at first regarded with suspicion as the work of 'left extremists'.[19]

Nygaard and others like him (including, famously, the founder of cybernetics, Norbert Wiener) believed that computers would have to come under democratic control or they would lead to disastrous imbalances in society. He involved the Norwegian Iron and Metal Union in discussions and, in the late 1960s, they established the right of workers to share control of the way new technology entered the workplace. This became national policy under the Data Agreement between unions and employers in 1975.

A tradition of 'participatory design' was established throughout Scandinavia where (in principle at least) workers guided the design of new technologies. Like object-oriented programming, these initiatives did not inevitably produce egalitarian results.

For example, one of the first 'desktop publishing' (DTP) systems emerged from a participatory design project aimed at 'work enrichment', under workers' control, at the offices of the Swedish newspaper *Aftonbladet* at the start of the 1980s.[20] Not long afterwards, DTP was helping newspaper owners to get rid of print staff and journalists in droves. The idea of working from home, which became such a very mixed blessing, started here, with the 'telecottage' movement – first in Sweden and then Japan – aimed originally at keeping rural communities alive and even promoting a population shift from towns to the country.[21]

These and similar initiatives became by imperceptible degrees part of the post-1980 drive, and especially after the rise of the

Web, to commodify the computer, gain competitive advantage and boost sales, under the rubric of 'usability'.

THE RETURN OF MEDIEVAL ECONOMICS

The 'great compression' ended during the 1970s, giving way to what economist Paul Krugman has called 'the great divergence' of incomes and wealth within and between nations, at around the same time as computers and digital electronics began to become a major force in economic life.

By the late 1970s, there was much talk in political and journalistic circles about 'the coming of the microchip' – the first of a wave of similar frenzies accompanying the arrival of the internet (and especially, the World Wide Web) in the 1990s, and social media and cloud computing two decades after that.

Computers and electronics have been blamed for the Great Divergence, and are certainly implicated in it. But other, explicitly anti-egalitarian forces were also at work. These included the warlike and anti-democratic policies of Richard Nixon (US President from 1969 until 1974, when he was forced to resign), and in particular the 'Nixon shock' of 1971: his decision to deal with the inflation caused by the US's war in Vietnam by allowing the dollar to 'float' against other currencies. This effectively demolished the post-War Bretton Woods system whereby global currencies had remained relatively stable, and plunged poorer countries into debt from which they have never recovered.

This deliberate lurch into inequality imposed its own timeless logic on the new technologies as they developed. But via them, it also had a huge impact on the world of work and consumption – as well as on the environment.

It has been argued that social inequality shapes societies according to a kind of built-in 'grammar', so that all unequal societies down the ages segregate people in the same kinds of ways, and develop similar social mechanisms for controlling them, with similar consequences.[22]

The scene today looks modern, and very different from even the recent past, if you consider only its ephemeral features (the fashions, the devices that people use, their job titles and so on). But if you look past them at the social structures underneath, and their environmental consequences, we are witnessing something very like a lurch back in time; not quite back to the Dark Ages, but certainly back to the 16th century and perhaps even to the 13th century.

Until the 17th and 18th centuries, the severest environmental impacts were confined to northwestern Europe and its immediate periphery. Today the same kinds and patterns of impact, and the oppressive social quirks that go with them, are observable globally.

Global inequality today is about the same as it was in the late Middle Ages in the first, tiny capitalist economies, in northern Italy and the Low Countries (modern Belgium and the Netherlands): around 0.7 on the Gini scale, which would award a score of zero to a society where wealth or income were distributed completely equally, and a score of 1.0 to a society where a single individual owned everything. The Gini index is a relatively blunt instrument, but many analysts of past inequality have used it, making it possible to compare a wide range of societies over time.

Various studies have found that this 'north-European norm' of 0.7 is without precedent in world history. No previous society had ever sustained anything like it for very long – not even ancient Rome[23] or Egypt.[24] But among the capitalist states that emerged in Europe from the 15th to the 19th centuries, coefficients of 0.7 or higher were common. A demographic historian, Guido Alfani, has worked out that:

> in Paris at the beginning of the 14th century, the Gini index of wealth inequality was about 0.7; in London, it was 0.7 in 1292 and 0.76 in 1319. [Then, in the 15th century, in Italy, the coefficient] for the cities of Pistoia, Pisa, Arezzo, Cortona, Prato, and Volterra ... was about 0.75. In Florence it was higher, reaching 0.785.[25]

In 1800, before European expansion had started in earnest, global inequality was insignificant. According to many historians, life was generally better for Chinese and Indian peasants than for their

British counterparts.²⁶ By 1820 (after the end of the Napoleonic wars) things were changing. The global Gini coefficient was somewhere around 0.50, and rising steadily. ²⁷ Branko Milanovic and Christopher Lakner have estimated that, by 2008, global income inequality was around 0.705, but possibly as high as 0.76, when adjusted for under-reporting of top incomes.²⁸ One could say that Boccaccio's Florence had taken over the world.

FOR ELECTRONICS, READ TEXTILES

The inequality embodied in something like an iPhone could not be more different from the egalitarianism that made it possible, but it is not without precedent. Whereas today's most insecure workforces survive by assembling electronic devices, in late medieval Europe they were producing fashion textiles. Rapid obsolescence and highly atomized, precarious workforces were and are characteristic of both industries.

By the 16th century, entire populations in Europe depended for their very existence on this kind of textile production, and fashion textiles had become an 'engine of growth' of greater significance even than agriculture. Their production was organized for maximum profit on the 'Verlag' or 'putting-out' system: what we now call 'outsourcing'.

All over Europe, merchant entrepreneurs were bypassing guild-based workers, who knew their trades from end to end, and employing networks of mainly rural, often largely female and/or juvenile workers, each specializing in some tiny part of the process but unable to comprehend let alone influence the course of the events on which her or his life came to depend. For the first time, there was 'no such thing as a job for life' for whole categories of the population. As now, livelihoods could and did vanish in a moment, with a change of fashion or of international prices, or the capitalist's discovery of another, cheaper source of labor elsewhere. Outsourcing was already the name of the game, and so was obsolescence.

It can now sound naive to suggest that the biggest profits from any product should be found at or near the place where it is produced (the world's richest countries might then be in Africa, for example). But that was always the case in traditional societies and the tendency is by no means passé. It still applies, very emphatically, to the work of top-echelon barristers, management consultants, surgeons, artists and financial advisers. It is still the case to some extent where labor still has strong negotiating power. But as one descends the gradients of power the reward-to-labor ratio goes into reverse and the big amounts migrate as far as they possibly can from the point of production.

The great French medievalist Fernand Braudel established that distance was 'a constant indicator of wealth and success' in the first capitalist economies.[29] In the Venetian sugar trade of the 15th century 'production was never the sector in which fortunes were made'.[30]

> [A] kilo of pepper, worth one or two grammes of silver at the point of production in the Indies, would fetch 10 to 14 grammes in Alexandria, 14 to 18 in Venice, and 20 to 30 in the consumer countries of Europe. Long-distance trade... was based on the price differences between two markets very far apart, with supply and demand in complete ignorance of each other.[31]

For comparison, at the time of the Foxconn suicides in 2010, workers making Apple iPhones and similar products received a basic wage of $130 per month: about one 31,000th of the annual salary of Apple's then-CEO, the late Steve Jobs (an estimated $48 million[32]). This was actually a relatively modest differential by electronics industry standards, because Foxconn and Apple were at least paying the minimum wage at the time, and Steve Jobs was only taking a nominal salary from Apple of one dollar a year (most of his income came from his shares in The Walt Disney Company).

In terms of *where* the bulk of the profit went, a Silicon Valley research company estimated in 2010 (by taking an iPhone 4 apart) that only $6.54 of its $600 retail price ended up at the point of assembly in Shenzhen: just over one per cent. The rest went to materials and components suppliers ($187.51), 'miscellaneous' ($45.95) and profit ($360, 60 per cent of the total).

> What the latest analysis shows is that the smallest part of Apple's costs are here in Shenzhen, where assembly-line workers snap together things like microchips from Germany and Korea, American-made chips that pull in Wi-Fi or cellphone signals, a touch-screen module from Taiwan and more than 100 other components.[33]

The components, whatever their nominal country of origin, are also made under a similar or harsher regime so, here too, a tiny fraction of their value ends up in the places where human labor is involved, which might be in one of the poorer regions of China, Indonesia, the Philippines or Mexico. To reiterate: this is the analysis for a relatively ethically produced product. The norm is much more extreme for other electronics products, and for other industries. An analyst of the global garment industry, Pietra Rivoli, commented to the authors of the article quoted above that 'the value goes to where the knowledge is': a dynamic that began with capitalism itself.

In contrast to the situation in the great empires of the East and South, the great fortunes of Europe's late medieval and early modern period were also made 'where the knowledge was', and power politics ensured that it remained as far from the point of production as possible.

KNOWLEDGE-BASED WEALTH

We talk today of 'the market economy' as if we mean a single thing that we all understand but, in his groundbreaking work on global economic history, Braudel pointed out that the term covers two very different things.[34] It refers to a market in the traditional sense, still found all over the world, where everyone can see everything that's on offer, who's offering it and what they want for it, and where terms, quality and sometimes even prices are regulated in a way that everyone understands. But it also refers to a capitalist market where the only people who have that information, and the means to get it, are the capitalists themselves. The name of the game is to get as much of that knowledge as possible, while keeping it from anyone else.

THE BLEEDING EDGE

Our conventional fiction is that market knowledge is available, instantly, to everyone, so that we can all immediately take our custom to whoever offers the best value, driving out inferior quality and boosting high quality as we do so. Braudel's examination of the ledger books and letters of the first great capitalists – the powerful Venetian, Florentine, German and Dutch merchants of the late Middle Ages – reveals what a fiction that has always been. These men were true entrepreneurs in the modern sense. They could be literally staking their entire wealth on the accuracy of the information they had; every scrap was valuable and guarded jealously (so these documents were kept in strongboxes, along with bullion and jewels). The biggest money was in overseas deals, where privileged knowledge could be exploited most, so merchants had good reasons for persuading the state to protect their trade and enforce compliance.

This panopticon-like 'knowledge economy' was unknown in many of the world's other advanced economies until much later and then introduced only through force (massive force, in India and China). The sociologist Maria Mies has described how, in Burma, markets were run by women when the British arrived in 1824. The markets were well-regulated, open affairs that couldn't easily be manipulated. The women who ran them also had their own parallel information networks and power structures that counterbalanced those of male society. This was apparently not macho enough for Imperial tastes. Britain's methods for bringing Burma into the modern world involved conscripting Burmese men and teaching them harsh, military virtues, and putting women in their place through a 'housewifization' policy.[35]

The 'knowledge-based inequality' described by Braudel only arises when a merchant elite can rely on having privileged access to market information. European merchants were the first ones who could build businesses, industries and economies this way. They, uniquely, could increasingly count on state support to enforce their positions, and this was possible because medieval European states were relatively small and weak, and therefore easily dominated by pressure groups. New technologies emerging at that time (especially firearms and ocean-going ships) allowed these hitherto insignificant states to launch themselves on the world stage.

The Belgian historian Eric Mielants argues that capitalism is precisely what happens when a merchant elite takes over a state's judicial and military apparatus, and deploys it to protect itself against risk, at the expense of other groups. Abroad, the state's ships could be increasingly relied on to protect and monopolize trade routes and, nearer to home, the state's power could be used to define who had access to justice and who didn't:

> *Burghers, for instance, could not be tried outside their own city and were not to be imprisoned outside city walls, nor could any non-citizen testify against a citizen.*[36]

The city became a 'power container', to use Mielants' term. 'Up to ca 1500 the city states were the "power containers" that made this... underdevelopment possible; after 1500, the emerging nation states performed this function.'[37] The local countryside became an 'exploitable periphery' from which surpluses were extracted up to and beyond the capacity of the people to sustain themselves, or of the land to renew itself.

This was done by direct taxes and the creation of indebtedness, as well as by monopolies and laws controlling labor. The modern parallels are striking. Another medievalist, Peter Spufford, pointed out in 1988 that this was a colonial relationship in miniature:

> *In the 1280s the countryside of Pistoia supported a tax burden six times as high as that paid by the city. This excessive tax burden was only part of the sucking-dry of the contado by the city at this period. In Tuscany, payments from the countryside to the city greatly exceeded payments from the city to the country. The countryside fell into a state of endemic debt to the city. Repeated and continuous loans from the city to the countryside, and the purchase of rent-charges on the countryside by city-dwellers, only made the situation worse. Contadini were compelled to concentrate on cash crops rather than their own needs, and a cycle of deprivation was set up not unlike that in parts of the Third World today.*[38]

THE BLEEDING EDGE

The high-profit industries, like textiles, were built on cheap, controllable labor, so border controls became an important part of the dynamic, initially between town and countryside. Passports of various kinds became prevalent from as early as the 12th century (in the form of confession certificates), along with distinctive clothing and badges for the least favored groups (Jews, Gypsies, the unemployed), detention, expulsion and even 'carrier's liability' (fines for those 'people traffickers' who tried to sneak vagrants into town in the backs of wagons or on barges). Controls intensified with the adoption of paper (a 'write-once' medium, not easily falsified) and the development of print technology in the 16th century. Johannes Gutenberg's early career was built on the manufacture of pilgrims' badges, and he used his press initially to mass-produce indulgence certificates (the lucrative 'cash alternative' to penance, against which Martin Luther was later on to rebel, in 1517). The proceeds funded production of his Bible in 1455.[39]

None of these restrictions applied to elites, who could go wherever they liked. Foreign elites were welcome within the charmed circle, insofar as they were helpful – just as the elites of ex-colonial countries are today, sending their children to Western universities, their money to Switzerland and attending the same elite events as their international peers.[40] This pattern was established at least by the early 12th century during the 'Northern Crusades' (the context of Eisenstein's film *Aleksandr Nevsky*), which had the economic effect of encouraging eastern-European dependence on supplying low-priced raw materials to the West. A historian of those events, Eric Christiansen, describes a very modern-sounding relationship between rulers and foreign merchants:

> [Rulers] patronized foreigners who could bring them wealth, information and military skills or entertainment, and tried to create conditions which would attract them... The Russian princes insisted that, if a Russian borrower went bankrupt, his foreign creditors were to get satisfaction first, and that rates of interest for long-term loans should be low, to suit the long-distance trader... This common pursuit of wealth was the privilege of a small class of specialists, the international trading elite.[41]

THE FIRST MODERN ENVIRONMENTAL CRISIS

Peripheralization led to remorseless depletion of natural resources. Two Belgian researchers, Katharina Lis and Hugo Soly, calculated that, by the beginning of the 14th century (the century of the Black Death), landowners were taking 50 per cent or more of peasants' output but only putting 5 per cent of their incomes back into the land:

> The nobility, who disposed of the necessary means to employ technological improvements, squandered the vast bulk of the extorted surplus, while the peasantry, for whom the deepening of capital was a question of life and death, could seldom accumulate sufficient reserves to take steps in that direction.[42]

Soil exhaustion became a general phenomenon and led to miserably low crop yields. Braudel found that:

> Wherever one looks, from the 15th to the 18th century, the results were disappointing. For every grain sown, the harvest was usually no more than five and sometimes less. As the grain required for the next sowing had to be deducted, four grains were therefore produced for every one sown.[43]

Nothing like this happened in the world's other economies. In the early 1900s, the deep, well-cultivated soils of China, Korea and Japan were a source of wonder to American agronomist F H King.[44] In the great empires of the East, which had had huge armies, strong peasantries and sophisticated bureaucracies throughout the Middle Ages, merchants were never able to commandeer the agrarian economy as they did in Europe. Instead, general welfare sometimes acquired a major share of state resources, as demonstrated for example by the huge system of canals (which had a large famine-prevention role) constructed over many centuries in imperial China, and the 'taccavi' system in India, whereby rulers had a duty to help peasants financially through difficult times.

Within the capitalist polities, subsistence seldom even registered as a possible cause for concern. During the famine and plague years of the mid-1300s, cities including Florence and Milan, in desperation, reduced the tax burden on the peasantry and even initiated half-hearted irrigation schemes, but by this time the damage was done; the peasants were deserting the land and nothing could stop them.[45] Immiseration and soil exhaustion gradually turned into a full-blown demographic and ecological crisis, experienced at the time, all over Europe, as a sudden exodus of impoverished peasants from the land and into towns. A dynamic had started that required constant expansion or collapse.

The difficulties became extreme during the 'little ice age' – the period from the early 16th till the early 19th centuries, when global temperatures fell by two whole degrees Celsius. Indeed, this has been blamed for Europe's crisis – but the climate challenge did not affect all areas as badly, if at all. Some places, like Japan (newly unified by the Tokugawa regime), came through it with both environment and people in very much better shape than at the beginning.[46]

In Europe, the demographic and environmental crisis had begun long before, and continued till long after. By the beginning of the 20th century people were abandoning Europe at a rate of nearly a million a year, making Europe's exodus the largest mass emigration in history.

AN UNEQUAL SOCIETY IS A DANGEROUS PLACE FOR POWERFUL IDEAS

This puts a different complexion on the traditional account of 'the rise of the West', which used to attribute Europe's sudden expansion to the particular mindset or social structures of Europeans, or even to their higher state of evolutionary development: a view that remained fairly dominant even within living memory.

Deborah Rogers, a Stanford University researcher and now Director of the Initiative for Equality (IfE), looked at the riddle of European expansion from a different angle: given that human societies have been so characteristically and strongly egalitarian over the long run of history, why did unequal ones (and extremely unequal ones at that) end up sweeping the field?

Her team used a computer-modelling approach, which revealed (contrary to expectation) that the takeover of egalitarian societies by unequal ones had nothing to do with superior leadership or organization. Instead, it appeared that unequal societies have a short-term competitive edge over more egalitarian ones because those who take the decisions do not have to suffer their consequences. Rogers writes: 'unequal societies are better able to survive resource shortages by sequestering mortality in the lower classes.'[47] In the longer term, however, these societies can only survive by proliferation – finding fresh populations to bear the burdens they create. She concludes:

> Unequal access to resources is destabilizing, and greatly raises the chance of group extinction... which creates incentive to migrate in search of further resources... leading to conquests of the more stable, egalitarian societies – exactly what we see as we look back in history.[48]

This fits with archeological evidence that, in the past, unequal societies were usually 'self-extinguishing'. If an unequal society is isolated from its neighbors by forests, mountains or seas, and lacks the technology or resources to overcome them, it exhausts its resources and collapses. But when these obstacles don't exist, or new technology becomes available that allows them to be overcome, neighboring societies can be made to 'share the pain'. When these also become destabilized, they can 'spill over' in their turn into adjacent lands, and so on.

By the 1400s, intensely unequal city-states and early nation-states in western Europe had already been 'sequestering mortality in the lower classes' for about four centuries (as evidenced by their skeletal remains). Large areas of Europe were on the verge of ecological and demographic collapse – but its elites were acquiring new technologies such as firearms and ocean-going ships that allowed them to reach and dominate societies that were further and further away, across mountains, seas and eventually oceans until, finally, the system became global.

Western expansion was the result of a deadly combination:

intense inequality, emerging just at a time when human technologies were evolving to a stage where they could become deadly weapons in the hands of unaccountable elites.

From the 11th century onwards, new technologies arrived on the scene like kerosene deliveries to a burning building.

WATER MILLS, AND HOW NEW TECHNOLOGY CAN BE A CURSE

The scene was set for Europe's descent into extreme inequality in the first centuries of the second millennium, and that period provides a clear example of what should have been good technology falling into the wrong hands, and becoming a curse.

Around the year 1000 there was a sudden proliferation of water mills for grinding corn, and this was examined in detail in 1935 by the historian who inspired Braudel, Marc Bloch.[49] Whereas there were only perhaps 200 water mills in England at the time of Alfred the Great (849-899) there were 5,624 at the time of the Domesday Book (1086). In northern France the pattern was similar: 14 mills on one river, the Aube, in the 11th century, 62 in the 12th and over 200 by the early 13th century.[50]

This looks like a thoroughly enlightened development and had always been presented as such, but Bloch showed that it was quite otherwise. Far from being generous gifts from kindly aristocrats, water mills were imposed on unwilling populations and used as instruments of extraction. Peasants were required to take all the corn they grew to the lord's water mill for grinding – so that the lord could help himself easily to whatever proportion he considered to be his just share: a much simpler arrangement than sending armed men around from household to household.

Hand mills were outlawed everywhere. Bloch found records of house-to-house searches for illicit mills in the 12th century, and at least one famous insurrection, at St Albans in 1274, which rumbled on until the Peasants' Revolt in 1381. All the water mills of that period for which records exist were imposed under feudal or monastic auspices.[51]

One sees just how all those great castles and abbeys came

into being. They rose almost literally from the ground as the mill-wheels turned, sucking wealth out of everything and everyone around them. It no longer seems surprising that one of France's most palatial châteaux, Chenonceaux, on the River Loire, began its life as a water mill.

It appears that hand milling has important nutritional advantages. Modern studies have found that the greater force and heat generated by mechanized milling can destroy nutrients, and more nutrients are lost if you have to grind your grain all in one go and then store it; over 80 per cent of its vitamins can be lost within three days.[52] In her book about Ladakh's peasants, *Ancient Futures*, Helena Norberg-Hodge says that hand milling may have explained the Ladakhis' excellent health, despite a diet consisting mainly of barley. Vitamin deficiencies started to appear after the introduction of petrol-powered mills.[53]

So perhaps the water mill's introduction even played a part in the decline in stature of ordinary north Europeans (described in the next chapter), which started at just this time.

FIREARMS TAKE A EUROPEAN TURN

There was sharp increase in European inequality between 1500 and 1650. Peasants' real incomes fell by as much as a half, and selective adoption of new technology played a part here as well. In a much-cited paper published in 1979, the economic historian John Pettengill demonstrated that a very large part of the decline could be explained by the widespread adoption of new, small firearms by the European nobility during these years.[54] Hand guns in 1500 were cumbersome items, fired by a glowing fuse or 'match'; useful in armies but not suitable for a casual demonstration of personal power. Leonardo da Vinci had invented a wheel-lock arrangement in the 1480s, which allowed a gun to be fired just by pressing the trigger; improved versions of this were spreading throughout Europe by the 1520s. Pistols became possible and horse riders could use them. By the mid-1600s cheaper, more efficient flintlocks were becoming common, but they were still far beyond the means of the peasantry.

The power of armed enforcers, militias, or just an individual with a gun, was multiplied with each improvement. Pettengill showed how, as these developments unfolded, rents and taxes on peasants rose by multiples; restrictions on the use of commons were introduced and enforced; peasant rebellions were crushed with greater and greater predictability.

The result in eastern Europe was a wholesale return to serfdom; in the west, the result was an increasing rate of migration from the land to towns (and in some cases, as in England, deportation of surplus population to new colonies in Ireland and then in America), and 'proletarianization', as merchant elites learned to harness the new, precarious population through the Verlag or 'putting out' system of manufacture.

Pettengill found similar increases in immiseration from Scandinavia to Italy, and from western Spain to Russia. He knew of 'only two exceptions to this pattern: Venice and the newly industrialized areas of Holland'.

Was this bound to happen, once the principle of firearms had been invented? Things didn't unfold in this way in other parts of the world. Firearms were well established in China by the 13th century but were not turned into personal weapons. They were adopted in the Arab, Persian and Indian empires by the 15th century, and developed there, but without the appearance of a culture of firearms like the European one. The historian Noel Perrin tells us that Japan acquired its first firearms, a pair of arquebuses, from a Portuguese trader in 1543 – the same year in which a French immigrant ironmaster gave England its first cast-iron cannon.[55] Japanese gun enthusiasts improved on these two originals at a furious rate and reduced the cost of a gun from 1,000 to just 2 gold *taels*. By the end of the 16th century, the Japanese were able to have the kinds of battles among themselves that the Europeans could not manage until the end of the 18th (one of these, the battle of Nagashino, is the centerpiece of Akira Kurosawa's film *Kagemusha, the Shadow Warrior*). They then 'de-invented' firearms during the long, peaceful period that followed the Tokugawa clan's final victory in 1600 and lasted until 1868. When the US Commander Perry arrived in 1855 no guns whatsoever were to be seen.

It happens that the Tokugawa period was also a period of great equality in Japan. As the demographic historian Osamu Saito describes it, it was a case of 'all poor but no paupers'.[56] Society was rigidly hierarchical and ferociously punitive but, as Braudel puts it, it 'bristled with "liberties" like the liberties of medieval Europe behind which one could barricade oneself'.[57] Compared with Europe, the gap between rich and poor was minuscule, reflecting a much more equal balance of power. Village self-organization was a force to reckon with. The samurai class (who became government employees after the Tokugawa ascendancy) were often only a little better off than the peasants who, unlike so many 'modernizing' European peasants, controlled their own rural industries and paid little tax on the skilled work they produced, and which became regarded as so essentially Japanese. Other historians note that the physical quality of housing and amenities improved immeasurably and 'when viewed through the lens of life expectancy, Japanese led surprisingly long lives'.[58]

The environment also recovered. In the earlier period, large parts of Japan had been stripped of timber (in large part for building castles, temples and monuments). Despite a doubling or even tripling of Japan's population between 1600 and 1700,[59] deforestation was permanently reversed from around 1670, apparently through community initiatives as well as government ones.[60] Japan still has a surprising amount of forest for such a densely populated country. In 1993, 63 per cent of its land area was covered in closed forest.[61]

From a purely functional and environmentalist point of view, Japan seems to support a saying of the epidemiologist and equality campaigner Richard Wilkinson that, to a certain extent, 'it doesn't matter how you get your greater equality, as long as you get it'.[62]

1   Stéphane Hessel, *Time for Outrage*, Quartet Books, 2010.
2   InterOccupy obituary for Hessel, nin.tl/InterOccupyonHessel Accessed 11 Aug 2014.
3   Neal Ascherson, review of Ian Buruma's Year Zero, *The Guardian*, 12 Oct 2013. See also Gerry Cordon, nin.tl/AschersonEurope Accessed 11 April, 2014.
4   Wilfred R Bion, *Experiences in groups, and other papers*, Basic Books, 1961.

5. Donald Woods Winnicott, *Playing and Reality*, Tavistock, 1971.
6. Theodor W Adorno, *The Authoritarian Personality*, Harper, 1950.
7. Stanley Milgram, *Obedience to Authority: an experimental view*, Harper & Row, 1974.
8. For a comprehensive update on these matters, see HM Ravven, *The Self Beyond Itself*, New Press, 2013.
9. Goldin & Margo, 'The Great Compression', *Quarterly Journal of Economics*, 107, Feb 1992, pp 1-34.
10. Timothy Noah, *The great divergence*, Bloomsbury, 2012.
11. Thomas Piketty, *Capital in the Twenty-First Century*, Harvard University Press, 2014, p 499.
12. Ibid, p 505, note 33.
13. Rob Rooke, 'A Brief Socialist History of the Automobile', Links, 19 May 2009, links.org.au/node/423
14. G Gorodetsky, ed, *The Maisky diaries: red ambassador to the Court of St James's, 1932-1943*. Yale University Press, 2015, p 475.
15. Caroline Moorehead, *Village of secrets: defying the Nazis in Vichy France*, Vintage, 2015.
16. Richard Wilkinson, *The Impact of Inequality*, Routledge, 2005, pp 208 & 302.
17. World Bank, *The East Asian miracle: economic growth and public policy*, Oxford University Press, 1993.
18. Lecture by Kristen Nygaard, nin.tl/Nygaardlecture
19. Kristen Nygaard, 'Those Were the Days'? Or 'Heroic Times Are Here Again'?, *Scandinavian Journal of Information Systems*, 1996, aisel.aisnet.org/sjis/vol8/iss2/6
20. Robert Howard, 'Utopia: where workers craft new technology', in *Perspectives on the Computer Revolution*, ed Liam Bannon & Zenon Pylyshin, Ablex, 1989.
21. Industrial Structure Council report, quoted by T Morris-Suzuki, Beyond Computopia: Information, Automation and Democracy in Japan, Kegan Paul, 1988, p 11.
22. Jim Sidanius & Felicia Pratto, *Social dominance: an intergroup theory of social hierarchy and oppression*, Cambridge University Press, 1999.
23. Vladimir Popov, 'Why the West Became Rich before China and Why China Has Been Catching Up with the West since 1949', New Economic School, Moscow, 2009, nin.tl/PopovChina
24. AY Abul-Magd, 'Wealth Distribution in an Ancient Egyptian Society', *Physical Review E*, vol 66, 2002.
25. G Alfani, 'Wealth inequalities and population dynamics in early modern northern Italy.' *Journal of Interdisciplinary History* xl, no 4, Spring 2010, 513–49
26. Prasannan Parthasarathi, *Why Europe grew rich and Asia did not: global economic divergence, 1600-1850*, Cambridge University Press, 2011.
27. D Held & A Kaya, *Global inequality: patterns and explanations*, Polity, 2007, p 84.
28. Christoph Lakner, Branko Milanovic, and World Bank, *Global Income Distribution From the Fall of the Berlin Wall to the Great Recession*, World Bank, 2013.

29 Fernand Braudel, *The Wheels of Commerce*, Collins, 1979, p 190.
30 Ibid, p 192.
31 Ibid, p 405.
32 'Steve Jobs thrives on dividend income', Seeking Alpha, Jan 2010, nin.tl/SteveJobsdivis
33 David Barboza, 'Supply chain for iPhone highlights costs in China', *New York Times*, 5 July 2010.
34 Fernand Braudel, *Afterthoughts on Material Civilization and Capitalism*, Johns Hopkins University Press, 1977, p 107; see also Braudel, *Wheels op cit*.
35 Maria Mies, *Patriarchy and Accumulation on a World Scale*, Zed, 1998.
36 Eric H Mielants, *The Origins of Capitalism and the 'Rise of the West'*, Temple University Press, 2007, p 155.
37 Ibid, p 160.
38 Peter Spufford, *Money and Its Use in Medieval Europe*, Cambridge University Press, 1988, p 247.
39 V Groebner, M Kyburz & J Peck, *Who are you?: identification, deception, and surveillance in early modern Europe*, Zone Books, 2007.
40 C Freeland, *Plutocrats: The New Golden Age*, Doubleday Canada, 2012.
41 E Christiansen, *The Northern Crusades*, Penguin, 1997, p 42.
42 C Lis & H Soly, *Poverty and Capitalism in Pre-Industrial Europe*, Harvester Press, 1982, p 28.
43 Fernand Braudel, *The Structure of Everyday Life*, University of California Press, 1992, p 120.
44 FH King, *Farmers of forty centuries; or, Permanent agriculture in China, Korea and Japan*, Mrs FH King, Madison, Wisconsin, 1911.
45 Spufford, op cit, pp 105-106.
46 Geoffrey Parker, *Global Crisis*, Yale University Press, 2013, pp 484-508.
47 Deborah S Rogers, Omkar Deshpande & Marcus W Feldman, 'The Spread of Inequality', *PloS One* 6, no 9, 2011.
48 Deborah Rogers, 'Inequality: why egalitarian societies died out', *New Scientist*, 30 July 2012.
49 Marc Bloch, 'The Advent and Triumph of The Water Mill', in J E Anderson (ed) *Land and Work in Medieval Europe: Selected Papers by Marc Bloch*, New York, 1969, pp 136-68.
50 Calum Roberts, *The Unnatural History of the Sea*, Shearwater Books/Island Press, 2007, p 23; citing RC Hoffman, 'Economic Development and aquatic ecosystems in mediaeval Europe', *American Historical Review*, 101:630-669.
51 Stephen A Marglin, 'What Do Bosses Do?', *The Review of Radical Political Economics* 6, no 2, Summer 1974, 60–112. p 104; citing Bloch/Anderson *Land and Work in Medieval Europe*, op cit.
52 A website selling modern hand mills for the kitchen tells us that 'Commercial milling removes nearly 30% of the most nutritious parts of the whole grain. Within 72 hours, whole grain flour has lost over 80% of most vitamins. Mold and bacteria also quickly combine to further reduce nutrients and taste. The wheatgerm oil quickly becomes rancid, leaving the flour tasting flat at first, and then bitter.' skippygrainmills.com.au/faq.htm

53 Ladakh Project: localfutures.org/ladakh-project
54 John S Pettengill, 'The Impact of Military Technology on European Income Distribution', *Journal of Interdisciplinary History* 10, no 2, 1979, 201, doi:10.2307/203334.
55 N Perrin, *Giving Up the Gun*, Nonpareil Books/ D R Godine, 1979.
56 O Saito, 'All Poor but No Paupers', Leverhulme Lectures, University of Cambridge, 2010.
57 Braudel, *Wheels of Commerce*, op cit, p 590.
58 WM Tsutsui, *A Companion to Japanese History*, Wiley, 2009; see also Susan Hanley, *Everyday Things in Premodern Japan*, University of California Press, 1999.
59 Parker, op cit, p xvi.
60 Gerald Marten, 'Japan – How Japan Saved Its Forests', *The EcoTipping Points Project,* June 2005, nin.tl/Japansilviculture
61 Marcus Colchester & Larry Lohmann, *The Struggle for Land and the Fate of the Forests*, World Rainforest Movement, 1993, p. 21.
62 Richard Wilkinson, various presentations including (online) 'The Levelling Spirit', *Communities in Control Conference*, Melbourne, 1 June 2010, nin.tl/Wilkinsonleveling

3

What inequality does to people

The evidence is incontrovertible: inequality damages human health. The poor suffer more from illness and disease; their height is reduced; and they die earlier. The traumas they suffer are even passed on to future generations. The time may come when policies promoting inequality, as under austerity, are prosecutable as crimes.

Societies have always tended to regard the poverty around and within them as if it were an unsightly growth that it ought to be possible to remove somehow, leaving the rest of society pristine and unaltered. But it now seems that the real specter at the feast is not poverty per se, but the inequality of the society as a whole.

Inequality is injurious in itself. It impedes and distorts technological progress. It also impairs people, makes them ill, causes them to die sooner – in very specific and increasingly predictable ways. The new evidence on this derives from breakthroughs in computing and statistical science that were achieved in the Second World War and its egalitarian aftermath. Computers made it possible to analyze health data for whole populations, and to discern patterns that had lain invisible before. In this way, the computer, and the skills that grew up with it, have served a similar revolutionary function to that of the microscope a hundred years earlier.

In the mid-1970s Richard Wilkinson, then a postgraduate researcher in community health, applied the new statistical and computer techniques to Britain's public health data and found that, while health overall had improved, unexpected, large, and growing differences in illnesses and life expectancy had appeared between the wealthiest and poorest fifths of the British population. The gap was two or three times as large in the 1970s as it had been in the 1930s.

When all other factors, such as local and cultural variations in diet and climate, had been eliminated, inequality stood out as the sole, constant predictor of increased child mortality, cardiovascular disease, stroke, various cancers... a list that would subsequently grow to include diabetes, obesity, mental illness, and a constellation of societal disorders, including youth suicide, alcoholism, drug use and violent crime. It wasn't just that illness was retreating more slowly among the poorest; new major illnesses, like diabetes, were emerging and claiming lives across the board, but especially among the poorest.

In December 1976, Wilkinson summarized his findings in an open letter to the British Labour government's health minister published in the magazine *New Society*. The health minister promptly commissioned a fuller study by an expert panel headed by the government's chief scientist, Douglas Black. Three years later, the Black Committee's report confirmed the trends Wilkinson had identified, and concluded that they were not just related to inequality – they were *caused* by inequality. The implications of this finding triggered other studies, and initiated a ferment of research and further revelations.

This turn of events might easily not have happened. The Black Committee delivered its report just weeks after Britain had elected one of the most aggressively anti-egalitarian governments ever chosen by a democracy – that of Margaret Thatcher – which tried to bury the report, failed, but turned Britain anyway into a nation-sized inequality lab.

Wilkinson retained his focus on the original anomaly: the failure of increased wealth to produce commensurate health benefits. His own work has continued to concentrate on rich countries.[1] By the 1990s, he was able to show that the trends he'd identified in the 1970s held true not only within other countries but also between them, and moreover that the trends waxed and waned in step with changes in economic policies – with a clear correlation between global pandemics, for example of depression and obesity, and the growth of economic inequality.

The findings imply that the health impacts of inequality can no longer be dismissed as just misfortunes. They arise not by accident but because of specific decisions and actions taken by specific individuals, who cannot escape responsibility for what happens.

INEQUALITY REDUCES LIFE EXPECTANCY

People of all social classes in highly unequal societies do not live as long as people in more egalitarian ones. In Britain (one of the world's most unequal rich countries), in the years immediately before the latest global economic crisis, being among the poorest 20 per cent shortened women's lives by 7 years and men's lives by 7.3 years.[2] In the world's most unequal large economy, the US, the penalty for being in that lower fifth was about 14 years.[3]

The financial crash wiped trillions off the world's economy, but it was those who were already suffering most who bore the brunt. Income and wealth inequality continued their upward trends and so did the death toll. By 2015 the British press was reporting official figures showing that 'People born in parts of the UK have lower life expectancy than those in war-torn Lebanon'.[4] Not only were people in the poorer areas dying younger; a much bigger chunk of those shortened lives was taken over by illness and disability. Also in 2015, researchers from the King's Fund reported a 17-year difference in 'healthy life expectancy' between the poorest and wealthiest health-service districts and 'if you happen to be female and live within the borders of NHS Guildford and Waverley in Surrey you will – on average – have 20 more years of healthy life than if you happen to be male and live within the borders of NHS Bradford'.[5]

Other harms that go with inequality – including depression, suicide and substance abuse – all follow the same pattern. So do the factors that might drive someone into despair: incidents of inter-personal violence, such as rape, are more common; so is vandalism. You are more likely to be imprisoned in a less equal society than in a more equal one: rates of incarceration reflect a society's position in the 'inequality league table'. Thus, the US prison population quadrupled between 1978 and 2005 – the main period during which inequality soared. At approximately the same time, Britain's prison population doubled, whereas it rose only modestly if at all elsewhere – and actually fell in Finland during the 1990s. Wherever the increases occurred, it was overwhelmingly the poorest and blackest who were imprisoned, which fits with another finding: that racist attacks are more common in more unequal societies.[6]

Those who defend inequality sometimes say it's needed because it rewards ambition; social mobility and equality of opportunity are what matter most, they say. Yet social mobility is *lower* in more unequal societies than in less unequal ones. Even the politically conservative journal *The Economist* has repeatedly focused on this phenomenon over the years[7] – and it is especially true in the 'land of opportunity' itself, the US.

Extending back in time, the statistics reveal that the trend toward greater inequality, and the harms that flow from it, began in earnest in the 1970s – in other words, Wilkinson 'hadn't seen nothing yet' when he wrote his article, and nor had Douglas Black's committee. This reversed a trend towards greater equality that had started just after the Great Depression (which started in the US in 1929, with the Wall Street crash, but slightly earlier in the UK, with reversion to the gold standard). In Britain, inequality had fallen most rapidly during the Second World War – and public health improved faster than at any other time in the country's history.[8] The gains were generally attributed to egalitarian measures, such as rationing, which ensured that many people had enough to eat for the first time, and everybody had a reasonably balanced diet, even if it was an extremely meager one by today's standards. There was a general turn away from ostentation. Large cars were taxed more heavily than small ones. The sense of solidarity that made those gains possible made an indelible impression on those who experienced it, with people of all classes recalling how 'we all pulled together during the War' – and it paved the way for the introduction of Britain's welfare state, the National Health Service, and the nationalization of major industries and utilities.

Wilkinson has collated figures that show similar phenomena in the countries of the pre-1990 communist bloc. Health in the Soviet Union improved steadily until the late 1960s, when reforms introduced greater pay differentials.[9] Growing inequality was mirrored by a deterioration in public health, and evidence of social breakdown and concerns from the older generation about growing selfishness, long before the Soviet system collapsed. Inequality then took off in earnest in the period known in the West as *perestroika* (restructuring) but in Russia as *katastroika*, when male life expectancy in particular collapsed – a period that, it was said, cost more Russian lives than the Second World War.[10] The one eastern European country where life expectancy continued to increase during the 1970s and 1980s was Albania – which was following its own independent Maoist line.

EQUALITY AND THE SOVIET UNION

Does this imply that, despite the general discrediting of the communist period and the unquestionable tyranny of Stalin's regime, the Soviet Union did at least tackle inequality? Egalitarianism was a huge social force in Russia before, during and after the Revolution, but Soviet leaders rarely gave it a high priority. Real equality was thought to be possible in an ideal communist world of the future, but not feasible in the here and now, and possibly a dangerous distraction that needed to be curbed, violently if necessary.

Writers such as Petr Kropotkin (the Russian aristocrat who became a founder of anarchist thought in the late 19th and early 20th centuries),[11] and more recently the US historian Richard Stites,[12] have documented the very strong, versatile and egalitarian mutual support systems that flourished among ordinary Russians – the 'artels' that people would form to pool wages, food, equipment, living space, in just about every situation imaginable; and the 'Mir' or *obshchina* system of collective land management in the countryside. Stites has described some of the innumerable egalitarian projects that flourished in the years immediately after the 1917 revolution.

THE BLEEDING EDGE

The Bolshevik leadership generally thought those traditions backward. Many of them were avid disciples of the 'scientific management' approach developed by Frederick Winslow Taylor, and applied with such impressive results in Henry Ford's factories. They believed in strong managerial control, top-down discipline and incentive systems. Under Stalin, egalitarianism became a dangerous deviance, to be stamped out – a major goal of the purges.

Incentives and 'perks' were always present, used at first mainly to keep politically unsympathetic but valuable experts on board. These could arouse forthright popular outrage, but distinctions became steadily entrenched during Stalin's rise to power in 1928. He launched the Great Terror of 1934-39 with a chilling speech, in which he ridiculed and denounced the 'levelling' tendency as 'leftist chatter'. Stites has described Stalin's impact:

> His hostility – voiced in sarcastic and dismissive terms – was so deep and so clearly enunciated that it rapidly became state policy and social doctrine. He believed in productive results, not through spontaneity or persuasion, but through force, hierarchy, reward, punishment, and above all differential wages. He applied this view to the whole of society. Stalin's anti-egalitarianism was not born of the five-year plan era. He was offended by the very notion and used contemptuous terms such as 'fashionable leftists', 'blockheads', 'petty bourgeois nonsense', and 'silly chatter', thus reducing the discussion to a sweeping dismissal of childish, unrealistic and unserious promoters of equality. The toughness of the delivery evoked laughter of approval from his audiences.[13]

The tough, scornful, 'realistic' tone of Stalin's speeches is not without its echoes in the capitalist world, where the same conviction prevails that 'you can't make an omelet without breaking eggs'.[14]

The 'apparatchik' culture was felt right through Soviet society, and was resented. Aleksandr Solzhenitsyn has described how, during some of the most desperate fighting in the Second World War, new insignia of rank and even Tsarist-style 'shoulderboards' reappeared in the army, and *The Internationale* ('rise up, you wretched of the earth!') was replaced as the Soviet anthem by the dirge-like ode to the 'homeland of Lenin', which became familiar at post-War Olympics. The 'party maximum wage' was formally abolished in 1932 and by the late 1930s members of the ruling elite were being paid 10 to 30 times what a worker could earn, plus enjoying free furniture and other perks.[15] In the 1970s the dissident physicist, Yuri Orlov, calculated that party heads were paid about 25 times the salary of an ordinary worker.[16] This was only slightly less than the pay gap in the US between chief executives and workers at that time (a 30-fold difference), but much less than it became in 2005 (110 times).[17]

Despite this onslaught, egalitarianism remained the bedrock of Soviet society, and it produced the sorts of outcomes we would expect: it outspent the UK fourfold on education per person in 1959.[18] The number of pupils in the top four grades of high school rose from 1.8 million in 1950/1 to 12.7 million in 1965/6.[19] The population recovered with astonishing rapidity from its multiple extreme traumas (the First World War, the Civil War and Wars of Intervention, collectivization, the purges, and then the Second World War).[20]

After Stalin's death Nikita Khrushchev tried to restore more egalitarian pay policies and even initiated one in which managers and foremen were paid *less* than ordinary workers.[21] This coincided with the upswing in public health mentioned above, which ran out of steam in the late 1960s, after Khrushchev was ousted.

The Soviet Union's greater equality comes down to two things. First, the whole edifice rested on a persistent and deep culture of solidarity and egalitarianism described by historians like Stites, and by veterans of the Soviet period. Second, the inequalities mentioned above were all ones of income, not wealth. Individuals could not accumulate the kinds of private fortunes enjoyed by elites elsewhere. Almost all of the country's assets were state property, so there was no legal way for individuals to exploit ownership and acquire rental income which, as Thomas Piketty has shown, is the main driver of extreme inequality, especially when allied to the possibility of inherited wealth.[22]

The impossibility of inheriting significant wealth in the Soviet Union meant that this was one of the few places on earth where, in a strictly economic sense at least, 'the living were worth more than the dead', to use Piketty's phrase.[23] This has hardly ever been the case in Western countries, apart from the 30-year post-War period known in France as 'Les Trente Glorieuses', during which 'for the first time in history... wealth accumulated in the lifetime of the living constituted the majority of all wealth: nearly 60 per cent.'[24]

In the Soviet Union, with rent and inheritance removed from the equation, people could put effort into welfare initiatives in a fairly direct way, bureaucracy permitting. Technology and science could sometimes explore important paths that were barred to Western firms by the logic of profit. In the West, for example, antibiotics became the standard treatment for bacterial infections (generating the profits necessary to the pharmaceutical companies that produced them, and which inevitably used their financial power to shape the course of research in directions that would ensure further profit). Soviet research flowed instead into the use of 'phage viruses' (viruses that invade and destroy bacteria). Phage viruses are not a consistent product that can be mass manufactured, stored and sold in bulk, or even patented – they are mainly found in soils, and need to be searched for and cultivated continuously to counter constant mutations in the target bacteria – so they are not attractive candidates for development within capitalist firms.[25]

Immediately private ownership of major assets became possible, in 1990, the hierarchies were there, ready to take advantage of the situation, unleashing the chaos that took conventional economists by surprise. Till the last, the Soviet Union had survived on the hope that egalitarianism would in the end prevail. The final crushing of that hope undoubtedly helps explain the scale of damage that followed *katastroika*.

A similar story unfolded, but a little differently, in China. In the 1970s (the period just after the Cultural Revolution and before Deng Xiaoping's marketizing reforms) wage differentials between party officials of different grades were already well above those in the Soviet Union. However, overall inequality was relatively modest (about the same as in Britain at the time, and enormously less than it had been before the Revolution), and China was deemed a 'high achiever' in terms of public health. Birth-rates, for example, were well on the way to stabilization before the brutal one-child policy was introduced in 1979.

After 1980, with the state-managed transition to a market economy (but retaining state ownership of land), inequality soared and health gains stagnated – despite a 12-fold increase in GDP. Health indicators are still rising for most people, but much more slowly than in the past, and rural, female and older populations have suffered particularly wretched reverses.[26]

AUTONOMY AND SOLIDARITY: THE ESSENTIAL NUTRIENTS

Before Richard Wilkinson's and then Douglas Black's reports in the 1970s, nobody had imagined there could be serious differences in health in populations, like those of the US and western Europe, where almost everyone enjoys what look like excellent standards of living in comparison to those of a century or even 50 years ago. How can this happen?

Much of the new health inequality turns out to derive from psycho-social factors, which are now known to be far more powerful than anyone had previously suspected.

Two famous studies (known as the Whitehall Studies) carried out over a number of years by epidemiologist Michael Marmot and his colleagues found that civil servants' job status was an extremely accurate predictor of their age at death. Those in lower grades did not live as long after retirement as their superiors did. In particular, subordinates had up to four times the death rates from coronary heart disease that their bosses did. This was found to be linked to higher levels of the blood protein fibrinogen, which is associated with the 'flight or fight' response in vertebrates: it helps blood-clotting and is released in situations where the animal feels threatened (and might suffer a wound leading to blood loss). Evolution did not anticipate the kinds of 'pecking orders' that are found in offices; as far as the body is concerned, a threat is a threat, and fibrinogen is released, even if the source of threat is a summons from a superior, or the thought of one.[27]

Whatever the cause of death or illness, the most significant factor was found to be lack of autonomy. Kate Pickett and Richard Wilkinson say: 'of all the factors that the Whitehall researchers have studied over the years, job stress and people's sense of control over their work seem to make the most difference'.[28]

Low status can be mitigated by group solidarity. In his 2005 book *The Impact of Inequality*, Richard Wilkinson tells the story of a village of poor, southern Italian immigrants who settled in the town of Roseto, near the US city of Philadelphia.[29] They remained far healthier than equally poor Americans who lived nearby, despite very poor diet and very high cigarette consumption, because they insulated themselves within their language and culture, maintaining their old sociability and egalitarianism, which included general disapproval of competitive displays of status (which they called 'putting on the dog'). But when a new generation grew up, speaking English, seeing themselves as Americans and comparing their own lives to the ones shown on TV, their health collapsed; they became normal poor Americans.

The downside of this protective, 'huddling' strategy is that an unequal society develops a class structure or caste system. People from lower-status groups become stigmatized. They are no longer seen as individuals but as examples of a disparaged sector of society. Members of higher-status groups become capable of almost any degree of cruelty towards them, while remaining in other respects decent, caring folk. This is the 'group inequality' phenomenon described by Heidi Ravven (Chapter 1), and which two other researchers, Mark Sidanius and Felicia Pratto, placed at the heart of the problems of unequal societies. Their book *Social Dominance* concludes that the multifarious forms of structural and physical violence that come with inequality are driven by:

> *a single, simple heuristic:* group inequality. *Like a fractal pattern observable from micro- to macro-levels of organization, group inequality is seen in psychological biases, in the effects of social*

contexts, in the biases of institutional discrimination, and, ultimately, in general social structure.[30] [emphasis in the original]

All oppressive societies, they argued, are built ultimately upon 'gender and age sets': in plain terms, the first groups to be denied full human status when inequality appears are women and children.[31]

INEQUALITY MAKES PEOPLE SHORTER

The effects of inequality on human beings are impressive, even to the naked eye. For one thing, the poor are always, on average, smaller than the rich. But culture is extraordinarily good at hiding the evidence in plain sight. The idea that the poor are tougher than the rich is so powerful that it can completely obliterate what stares you in the face: that most of them are in much worse shape physically.

Sir William Hayter, who became Britain's ambassador to Moscow in the 1950s, has described the occasion in the early 1930s when the scales fell from his own eyes. He was no sportsman, hated athletics, and his university friends were of the same persuasion. Nonetheless, the father of one of them, Sir Robert Witt, arranged for them to take part in a tug-of-war against the locals one public holiday.

> We were all of us more or less unathletic, and the opposing teams, mainly young men from Brighton, had been practising all day. We expected to be ignominiously defeated. On the contrary, hardly had we touched the rope than our opponents practically fell into our arms. Sir Robert reduced our team, enlarged theirs, but nothing we could do enabled us to lose plausibly. Eventually we shambled back into the Old Clergy House, feeling that Sir Robert's well-meant attempt to create good will hadn't really worked out quite right.[32]

The physical differences between rich and poor are normally so well wrapped-up in cultural baggage that they can be ignored for millennia, and have been. Just occasionally we find flashes of recognition like the one above, or Captain John Smith's famous comment (from the early 1600s, on encountering a society that did not have rich or poor) that 'the Indians seemed like Giants to the English'.[33]

It is only since archeologists have started systematically measuring skeletons that the scale and nature of the discrepancy have become apparent.

The archeologist Martin Jones locates the divergence between rich and poor at around 5,000 years ago, in the early aristocratic societies of the eastern Mediterranean:

> the ordinary people have five times more dental lesions than their ruler and are up to four per cent shorter. An average Bronze Age male farmer from the eastern Mediterranean would stand 167 cm (five feet six inches); 6 cm shorter than his ruler and 10cm shorter than his hunting ancestors.[34]

The relatively new science of 'anthropometry', which uses skeletal data, military, hospital and workhouse records and the records of slave owners, among other sources, is making the story clearer. The onset of the really intense, European inequality described in Chapter 2 correlates with the anthropometric data.

Richard Steckel has found that, until around the 10th century, north Europeans were about as tall as they are today, but 'northern European men had lost an average 2.5 inches of height by the 1700s, a loss that was not fully recovered until the first half of the 20th century'.[35] Some of the fall in stature might have had something to do with medieval climate change, he says, but most of it is down to inequality: the changes in stature occur in places that did not experience climate change as well as in ones that did. Stature seems to be exquisitely sensitive to inequality – as illustrated by another study he mentions, showing that the children of modern British unemployed working-class men are shorter than those whose fathers have jobs.

Roderick Floud, Kenneth Wachter and Annabel Gregory have examined British health and stature in the 18th century (the period of Britain's rise to global power). They studied the records for

Royal Marine recruits of Nelson's navy, who were usually supplied by a pauper charity, the Marine Society. They write that:

> The boys of the Marine Society were extraordinarily short, particularly in the 18th century. Thirteen-year-olds born in 1753-80 average 51.4 in (130.6cm), a full 10 inches (25.4cm) less than the children of London measured by Tanner and others in the 1960s. The full contrast, in both the 18th and the early 19th centuries, is brought out vividly by [plotting] two birth cohorts against the modern British standard growth chart; if a Marine Society boy of 1787 were miraculously transported into a doctor's surgery in 1987, his next step would be into hospital as a sufferer from undernutrition or child abuse.[36]

They also describe the 'the dip [in stature] for those recruited during the "Great Immiseration" of the turn of the 19th century, the rise after the Treaty of Vienna, and the dip for those recruited after the mid-1840s' – giving the lie to the traditional idea that the transition to industrialism was a wonderful thing for Britain's poor. It is hard to argue with skeletons.

From the mid-19th century onwards we find increasing rumblings of concern about the physiques of recruits in armed forces in Europe, and alarm in Britain in the 1890s at the state of the young but toothless British recruits who were expected to fight well-built Boer farmers in South Africa. A recent history of Britain says that, by 1914, the average British soldier was five inches (12.7 cm) shorter than his officer.[37]

Only in the last generation or two have typical Europeans regained some of their earlier stature and health, but very patchily. Danny Dorling and his colleagues found differences in stature and dental health almost identical to the Bronze Age ones given above, in Sheffield in 2009 – between the poorer district of Brightside, and the neighboring, wealthier one of Hallam (and this is in one of Britain's less unequal cities).[38]

European Americans regained their lost stature in the 18th century, thanks to the treasure trove of high-yielding food plants developed over centuries by the American peoples – squashes, potatoes and especially maize[39] – but in Europe itself the recovery had to wait for the arrival of the technology that made the great wars of the 20th century possible: the Haber process for fixing nitrates (first as fertilizers, then as explosives). Larry Lohman and Nicholas Hildyard have written that 'the coal-intensive Bosch-Haber fertilizer-manufacturing process that tripled crop yields during the 20th century now accounts for half the nitrogen in every human body'.[40]

It is as if, having run through the possibilities of exploiting other lands, science and technology have given the world's powerful, unequal societies the keys to the planetary past as well, so that it too can be plundered via the exploitation of fossil fuels, whereby 400 years' worth of accumulated, ancient plant growth can be burned every year.[41]

TODAY'S INEQUALITY WILL DAMAGE FUTURE GENERATIONS

We are now learning that inequality doesn't only affect the generation that experiences it; it also affects their children and grandchildren, even after life has improved.

The first glimmerings of this sobering fact came in the Soviet Union in the 1960s, when the geneticist Raissa Berg and her colleagues carried out a statistical analysis of birth defects over previous decades. The more data they accumulated, from hospitals, from mortuary records and collective-farm ledgers, the clearer was the evidence that the Soviet people had suffered two terrible shocks during the 20th century, resulting in two 'spikes' in the numbers of birth defects.

The first of these spikes covered the late 1930s and early 1940s. Berg realized that the mothers of these children had all been born during the last, harsh year of the First World War and the famines that accompanied the revolution, counter-revolution and Wars of Intervention, and in the same areas of the Soviet Union that had suffered most. The second spike had occurred in the late 1950s: too early to be explained by the suffering in the Second World War (which did not hit the USSR till 1941) and rooted instead in

1936-38, the years of Stalin's 'Great Terror'.

The mothers of both generations of damaged children had developed in their own mothers' wombs under fearful conditions, sharing the terror and privation as they developed and their own ovaries formed, complete with the egg-cells from which their own damaged babies would grow. The two 'spikes' of birth defects pointed straight at these huge events, the second of which had otherwise been erased from history.

Similar stories are now emerging in the capitalist countries. The suffering inflicted on the people of northern Britain during the Great Depression of the late 1920s and 1930s was assumed to be over and done with, and eliminated completely by the welfare state. Wilkinson's work and the Black Report began to shake that assumption. By the early 2000s – despite vast improvements since the 1940s in healthcare, nutrition, education, working and living conditions – mortality rates and levels of heart disease and diabetes in the north of England and in Scotland were still well above those of the south. David Barker and his colleagues at the University of Southampton have discovered that the anomalies result from genetic damage suffered by those who survived the Depression in the places where it was most severe.[42]

Other studies, done in the Netherlands (which suffered an intense famine in 1944-5 under Nazi blockade), in Scandinavia (parts of which suffered alternating famine years at the turn of the 20th century) and in China (among survivors of the famines that accompanied Mao's 'Great Leap Forward') have revealed that the damage is carried down the generations by men, as well as by women.

Kent Thornburg has studied the effects in the southern US, where black and poor white populations still bear the genetic damage caused by the famines and violence that followed the Civil War of 1861-5 – damage reinforced by a further wave of hunger and societal collapse in the 1930s. Thornburg says that heart disease, stroke and diabetes are 'still rampant' in the Old South. 'Their malnutrition goes back to the Civil War and it has never gotten back on track.'[43] For the black population, it goes back even further; those waves of damage came on the back of the greatest trauma of them all, enslavement.

Inequality is a kind of violence – and in no way a metaphorical one. It carries the same level of human cost as any other kind of injury; it also has specific, human causes – actual individuals have argued, worked for, condoned and gone along with the policies that caused the injuries, and bear responsibility as with any other harmful act or failure to act. Ignorance of the facts could soon be no more a defence for politicians and civil servants, than ignorance of the fire regulations is for a factory boss. The debate is no longer 'just' about rights, but about negligence and abuse of power.

Could tolerating inequality become as unfashionable as smoking? Could it some day be viewed with the same automatic revulsion as tolerating wife beating, food adulteration or toxic pollution – all of which are in fact more common in more unequal societies than in more equal ones?[44] In the past, politicians knew none of this, but today's politicians can't plead ignorance. How long before 'austerity' enthusiasts feel a social chill around them, or even face class actions in court?

1 See Richard Wilkinson & Kate Pickett, *The Spirit Level*, Penguin, 2009; Richard Wilkinson, *The Impact of Inequality*, Routledge, 2005.
2 C White et al,. Trends in life expectancy by social class 1972-2005. *Health Statistics Quarterly*, 36, 2007. Available at: nin.tl/lifeexpecttrends
3 Vicente Navarro, 'Inequalities Are Unhealthy'; *Monthly Review,* vol 56, no 2, monthlyreview.org/0604navarro.htm
4 Andrew Gregory, 'People born in parts of UK have lower life-expectancy than those in war-torn Lebanon', *Daily Mirror*, 4 Nov 2015.
5 David Buck & David Maguire, Inequalities in life expectancy, King's Fund, August 2015.
6 Wilkinson and Pickett, op cit, 2009.
7 'Ever higher society, ever harder to ascend', *The Economist,* 29 Dec 2004; 'Nomenocracy', *The Economist*, 9 Feb 2013.
8 Amartya Sen, *Development as Freedom*, Knopf, 1999.
9 Wilkinson 2005, op cit, pp 114-116.
10 This seems to be an overstatement, but only a fractional one. Nicholas Eberstadt calculates that around 23.8 million lives were lost in the Russian Republic in the 16 years that followed the breakup of the USSR: 13.7 million fewer births, and 10.1 million more deaths than in the preceding 16-year period. Nicholas Eberstadt, 'Drunken Nation – Russia's Depopulation Bomb', *World Affairs,* 138, no 4, 2009, p 51. RW Davies has calculated that 38 to 39 million lives were lost during the Second World War: *Soviet Economic Development from Lenin to Khrushchev*, Cambridge University Press, 1998, p 2.

11 Petr Alekseevich Kropotkin, *Mutual Aid: A Factor of Evolution*, 1902.
12 Richard Stites, *Revolutionary Dreams*, Oxford University Press, 1989.
13 Ibid, p 232.
14 The Romanian communist and writer Panaït Istrati was told this when he visited Petrograd in 1917, and was arrested after he replied: 'OK. I can see the broken eggs. Where's the omelet?' Cited in Sébastien Lapaque, 'Panaït Istrati, roi des vagabonds', *Le Monde Diplomatique*, July 2015.
15 Stites, op cit, p 233.
16 Chrystia Freeland, *Plutocrats*, Doubleday Canada, 2013, p 90.
17 Ibid, p 135
18 Stafford Beer, *Decision and Control*, Wiley, 1966, p 478, citing UK government report 'The long term demand for scientific manpower', 1961.
19 RW Davies, op cit.
20 According to Davies's calculations, Stalin's collectivization, forced industrialization and Great Terror caused 10 million excess deaths. The First World War and its aftermath caused 16 million excess deaths, and a deficit of 10 million births.
21 Francis Spufford, *Red Plenty*, Graywolf Press, 2012.
22 Thomas Piketty, *Capital in the Twenty-First Century*, Harvard University Press, 2014, p 531.
23 Ibid, p 396.
24 Ibid, p 402.
25 Jean-Paul Pirnay et al, 'Introducing yesterday's phage therapy in today's medicine', *Future Virology*, 7(4), 2012, pp 379-390.
26 Steef Baeten, Tom Van Ourti & Eddy van Doorslaer, 'Rising inequalities in income and health in China', *Journal of Health Economics*, vol 32 (6), 2013, pp 1214-1229.
27 Wilkinson, 2005, op cit, p 163.
28 Wilkinson & Pickett, op cit, pp 75-76.
29 Wilkinson, 2005, op cit.
30 Jim Sidanius & Felicia Pratto, *Social Dominance*; Cambridge University Press, 1999, pp 304-5.
31 Ibid, p 199.
32 William Hayter, *A Double Life*, Hamilton, 1974, pp 17-18.
33 Karen Ordahl Kupperman, *Indians and English: Facing Off in Early America*, Cornell University Press, 2000, p 46.
34 Martin Jones, *Feast: Why Humans Share Food*, Oxford University Press, 2007, p 248.
35 Richard Steckel, 'New Light on the "Dark Ages"', *Social Science History* 28, 2, pp 211-229, 1 July 2004.
36 Roderick Floud et al, *Height, Health and History: Nutritional Status in the United Kingdom 1750-1980*, Cambridge University Press, 2006, pp165-166.
37 Andrew Marr, *A History of Modern Britain*, Macmillan 2009.
38 Danny Dorling, Personal communication, 6 Jan 2010.
39 J Weatherford, *Native Roots: How The Indians Enriched America*, Fawcett Columbine, 1991.
40 Larry Lohmann & Nicholas Hildyard, *Energy, Work and Finance*,

The Corner House, 31 Mar 2014, p 58, nin.tl/cornerhouseEWF Their source is Andrew Nikiforuk, *The Energy of Slaves*, Greystone, Vancouver, 2012, pp 79-81.
41 Lohmann & Hildyard, op cit, p 38.
42 GJ Burton, DJP Barker, A Moffett & K Thornburg, *The Placenta and Human Developmental Programming*, Cambridge University Press, 2010.
43 BBC Radio 4, 'The First 1,000 Days: A Legacy for Life, In the Womb', BBC, Aug 2011, bbc.co.uk/programmes/b0137z06.
44 JK Boyce, 'Is Inequality Bad for the Environment?', no 135, 2007, nin.tl/Boyceinequality

4

The environmental cost of human inequality

The more unequal a society is, the more it damages its environment, which eventually collapses. This is clear from archeological as well as modern evidence. However, more equal societies can enjoy prosperous and fulfilling lives with little or no impact. 'Blame games' have dominated the debate until now – targeting the rich, or the poor, or human nature itself. But our real enemy is the inequality that creates rich and poor – and turns technological progress into an arms race.

To a conventional 'hard-headed realist', the human suffering described in the last chapter is water off a duck's back. The idea that life is a tough business has been thoroughly drummed into many people. But what if all this allegedly healthy rough-and-tumble is what's wrecking the planet?

We have seen that inequality is bad for technological progress and bad for human beings. But if it is also, by exactly the same token, disastrous for the environment, then human well-being suddenly becomes a matter of major concern even for 'hard headed realists'. A simple if challenging remedy offers itself as an alternative to all the hand-wringing, finger-pointing and head-scratching: to reduce impact, reduce inequality.

ARE THE RICH DESTROYING THE EARTH?

It used to be argued that, although the wealth of the rich might be morally objectionable, it is a distraction because no matter how wealthy they are, sharing their wealth among the whole population would spread it too thin to make much difference. Even some leftwing economists have argued along those lines.[1]

But this misses the real point: when a few people control most of the wealth, the whole dynamics of a society change, affecting the way people live their lives, and the impact they collectively have on their environment.

We now have strong evidence from a great many sources that the lion's share of any society's impact is caused by its wealthiest members, and a less equal society *as a whole* has a higher environmental impact than a more equal one does.

An Oxford University study in 2006 found that 61 per cent of all travel emissions came from individuals in the top 20 per cent, while only 1 per cent of emissions came from those in the bottom 20 per cent.[2] Similar figures are frequently cited for the global situation. The social geographer Danny Dorling reckons that:

> it is almost certainly an underestimate to claim that the richest 10th of the world's population have a greater negative environmental impact than all the rest put together... And, of the richest 10th of the world's population, the richest 10th [the top 1 per cent] consume more, even than the other half a billion or so affluent.[3]

Their disproportionate impact is explained not by their wealth per se, but their wealth relative to the rest of the population. In fact, the very meaning of wealth is distorted in an unequal society.

Data from the World Wide Fund for Nature (WWF) in its biennial Living Planet Reports support the idea that global environmental impact is connected with inequality. The WWF developed the concept of the 'Global Hectare' (GHa) to measure how much of the planet each country requires to sustain itself at its current level of consumption. To be sustainable, each citizen would need no more than 2.1 GHa but, in 2005, US citizens were using 9.4 GHa each.[4]

This could easily be taken to mean that improved living standards inevitably carry an environmental cost, but the WWF's 2006 Living Planet Report revealed an important new twist by comparing each country's ecological impact with its quality of life, as measured by the UN's Human Development Index (HDI).

The environmental cost of human inequality

It showed that many countries with acceptable HDIs had much lower impacts than the US (and we will see more detail on this below). One of these, Cuba, made headlines briefly in the world's media: it was the only country in the world to achieve environmental sustainability and good lives for its people, with a footprint per citizen of just 1.8 GHa. It was even regenerating forests that had been destroyed in the days of Columbus. Cuba is of course (or at least still was in 2006) one of the world's most equal countries.

Cuba's huge achievements were recorded in the immediate aftermath of what should have been a catastrophe: the loss of its major export market and its main source of oil, fertilizer, medical technology, training, drugs and even soaps and disinfectants, after the collapse of the Soviet Union in 1989. Brian Pollitt, a Glasgow academic who works with Scottish Medical Aid for Cuba, has written:

> *The most important single factor is that despite dramatic falls in State income of all kinds during the worst years of the 1990s, the maintenance of health provision, together with education, was an official national priority... Cuban State expenditure on health was $937.4 million in 1990 and $1,210 million in 1996. In telling contrast, expenditure on defence and public order over the same years fell from $1,149 million to $725 million.*[5]

The fall in military spending is indeed telling – in at least three ways. First, military activities account for *a fifth* of all global environmental degradation[6] – so this of itself must have helped Cuba 'go green'. Second, militaries are quintessentially elitist organizations, and accrete around themselves larger 'penumbras' of important people, who supply, advise, admire and represent them in government, and this affects a country's respect system as well as its distribution of resources. Third, to reduce one's armed forces so significantly at such a moment of extraordinary vulnerability – Cuba had lost its main military ally and was now surviving, under blockade, on the very doorstep of its avowed enemy, which happened to be the world's largest military power – flies in the face of conventional wisdom, and rather calls it into question.

For those who are wary of accepting a Latin American socialist example, Finland also offers a similar example of very high health and educational achievements, low spending on military, police and prisons, very low carbon emissions – and very low inequality. Pasi Sahlberg has explained that Finland's politics were shaped by two major economic crises – in the 1970s and again in the 1990s – leading to an all-party commitment to solidaristic policies.[7] He writes: 'a crisis can spark the survival spirit that leads to better solutions to acute problems than a normal situation would.' These included scrapping selection in schools, abolishing private education of all forms and making it illegal, and a commitment to free education, including tertiary education, for everyone, wherever they lived. It was also decided to give teachers much greater autonomy, and scrap the schools inspectorate – of which Sahlberg had been Director. He instead became Finland's proselytizing education ambassador.[8]

Some people argue: 'Ah, but Finland is a rich country.' Not so: it has few natural resources apart from timber and it has never had an empire; the Finns had realized that their main resources had to be themselves and their children. The other objection is 'Ah, but the Finns are ethnically homogeneous: solidarity comes easily to them.' Again, not so. Finland is fairly diverse. It has several official languages, a substantial aboriginal population – the traditionally stigmatized Sami – plus Swedes, Russians, Romani, Arabic-speakers, Vietnamese, and others.

Measuring the impact of capitalist nations is likely to under-report the situation if one looks only at what is happening within their national borders, because their basic dynamic involves externalizing high-impact activities. But, even so, general patterns are observable – and the figures in the table below, based on the polite convention that nations do their own dirty work, are certainly on the conservative side.

The WWF's reports consistently show that countries, such as the Scandinavian ones and Japan, that are almost as wealthy in GDP per capita terms as the US, but which are more equal, have lower ecological impacts. Danny Dorling has found that the

The environmental cost of human inequality

correlation between European countries' inequality and their ecological footprints may be even closer than the reports suggest, when each country's usable land area is added into the equation.

Table 1 shows the 'eco-credit' (or debt) of 24 countries, and their Gini coefficients of income inequality, where available. It shows the impact of the country as a whole (1A) and per capita (1B). Nearly all the countries that are in 'eco-credit' are the more equal ones – especially when calculated per capita. It is interesting that some of the most resource-hungry countries are also the more secretive ones, with no readily accessible wealth and income data.

Table 1 National and per-capita ecological credit or deficit

A: National footprint			B: Per capita footprint		
Eco-credit	Country	Gini	Eco-credit	Country	Gini
37.97	Sweden	27.3	80.39	Iceland	26.9
28.18	Finland	27.1	5.42	Finland	27.1
24.11	Iceland	26.9	4.27	Sweden	27.3
4.90	Norway	25.9	1.09	Norway	25.9
1.61	Ireland	32.5	0.41	Ireland	32.5
0	Holy See	?	–0.14	Austria	30.5
–0.12	San Marino	?	–0.90	Denmark	29.1
–0.13	Liechtenstein	?	–2.39	France	33.1
–0.17	Monaco	?	–2.55	Portugal	35.0
–0.22	Andorra	?	–2.63	Germany	30.1
–1.11	Luxembourg	34.8	–2.77	Luxembourg	34.8
–1.26	Malta	?	–2.85	Italy	35.2
–9.21	Austria	30.5	–3.05	Switzerland	31.6
–10.24	Denmark	29.1	–3.15	Malta	?
–21.94	Switzerland	31.6	–3.19	Greece	36.7
–25.46	Portugal	36.0	–3.22	Spain	35.9
–35.08	Greece	36.7	–3.26	Andorra	?
–41.25	Belgium	27.6	–3.69	The Netherlands	28.0
–59.39	The Netherlands	28.0	–3.80	Liechtenstein	?
–132.19	Spain	35.9	–4.01	Belgium	27.6
–142.66	France	33.1	–4.02	United Kingdom	32.6
–163.68	Italy	35.2	–4.47	San Marino	?
–216.54	Germany	30.1	–4.92	Holy See	?
–237.70	United Kingdom	32.6	–5.02	Monaco	?

Eco-credit or deficit is shown in millions of global hectares for a selection of European countries, adjusted from WWF 2006 data by Danny Dorling to include each country's own biocapacity (usable land area). The unshaded countries at the top have more capacity than they are using, especially when calculated per capita (B).

The Gini coefficients are from the World Bank and are based on 0 = perfect equality and 100 = perfect inequality.

A similar relationship can be seen at regional and local level. Jim Boyce, co-director of the Political Economy Research Institute (PERI) at the University of Massachusetts, has found that the most unequal US states (such as Tennessee, Alabama and Mississippi) contain more environmental degradation than do more egalitarian states (such as Minnesota, Maine and Wisconsin), in direct relation to their level of inequality. The relationship holds up whether you measure this by income differences, or by distribution of political power – indicated by voter participation, access to education and healthcare, and tax fairness. This is because:

> In practice, many environmental costs are localized, rather than being uniformly distributed across space. This makes it possible for those who are relatively wealthy and powerful to distance themselves from environmental harm caused by economic activities.... Within a metropolitan area, for example, the wealthy can afford to live in neighborhoods with cleaner air and more environmental amenities. Furthermore, sometimes there are private substitutes for public environmental quality. In urban India, for instance, where public water supplies are often contaminated, the upper and middle classes can afford to consume bottled water. The poor cannot. In such cases, because access to private substitutes is based on ability to pay, again the rich are better able to avoid environmental harm.[9]

He, too, finds that the same phenomenon is observable globally: countries with a more equitable balance of power (evidenced by such things as civil rights and adult literacy) also tend to have better environmental quality – even when controlling for factors like per-capita income, which might have affected the result. He concludes that 'People are not like pondweed. How we treat the natural environment depends on how we treat each other'.[10]

Human environmental impacts have increased over time, as human inequality has increased. Figures from the International Panel on Climate Change (IPCC) show that atmospheric CO_2 equivalents increased more than twice as fast during 1995-2004 (the first 10 years of the World Trade Organization's existence, when the brakes really came off neoliberal growth and worldwide inequality soared) as during 1970-94.[11]

INEQUALITY TURNS HUMANS INTO A GEOLOGICAL FORCE

In 1992, a science writer and ecologist, Andy Revkin, wrote: 'We are entering an age that might someday be referred to as, say, the Anthropocene. After all, it is a geological age of our own making.'[12]

In 2008, the Geological Society of London's Stratigraphy Commission (the global authority on these matters) agreed to adopt Revkin's suggestion and make the Anthropocene designation official.[13] The Holocene period (the period of climate stability since the end of the last ice age, about 10,000 years ago) is officially over and we are now in 'a stratigraphical interval without close parallel in the last several million years'.

But this situation has been longer in the making than we generally realize. Archeology tells us that the problems of ecological collapse we're now facing were already evident when the first class-based societies appeared in the eastern Mediterranean around the third millennium BCE, and in the Americas. Those vanished societies should be 'warnings from history': their inequality made them 'self-extinguishing', for reasons explained by Deborah Rogers' work, described in Chapter 2. The heavy material demands that social inequality creates (for palaces, temples and so on) eventually sucked the life out of the ecosystems on which they depended, they were unable for one reason or another to draw in new resources from beyond their borders, and they collapsed. This was the fate of many of the states of early antiquity in the Fertile Crescent in present-day Iraq and Syria, and of the Mayan empire of pre-Columbian America in the ninth century CE. The events described in Chapter 2 suggest that Japan may have escaped a similar fate in 1600.

The Mayan empire fell victim to droughts caused by deforestation which, it is now apparent, was caused by the huge demand for charcoal to burn lime, required for the construction of palaces and temples. Thomas Sever, co-author of a recent study of the Mayan deforestation, calculated that 20 trees would have been needed to produce a square meter of cityscape. 'When you look at these cities and see all the lime and lime plaster, you understand why they needed to cut down the trees to keep their society going.'[14] The environmental effects are mirrored in the skeletal remains of the people, which show the kinds of malformation found in unequal societies down the ages.

As we go back in history, it becomes clear that the impacts of inequality on the environment and on humans cannot be viewed separately: they are a single system. The condition of one tells us about the condition of the other.

The archeologist Martin Jones represents a growing body of research indicating that human societies started to damage their environments precisely when and where they first started to become unequal, in Bronze Age settlements in the eastern Mediterranean, around 5,000 years ago. These contain typically 'modern' contrasts: the treasure-laden burials of tall, straight-limbed elites, alongside the sparser evidence of dwarfish, ill-clad, prematurely aged, malnourished subject classes, and rectangular fields stripped of nutrients by the insatiable demands of the elite for tradable commodity crops.[15]

Jones also reports that in Britain, excavations of graveyards from Roman times right up to the 20th century have revealed the skeletal malformations typical of a people subsisting on the most basic and monotonous of commodity crops, alongside skeletons with the spinal malformations typical of an elite grown obese on the meat raised and fattened on those commodity crops.[16]

MALTHUS'S MISTAKE: NOT TOO MANY BABIES, BUT TOO MUCH DEBT

Human and environmental degradation have been and still are widely attributed to population growth, and the introduction of agriculture. The influential science writer Jared Diamond has called agriculture 'the worst mistake in the history of the human race'. [17] The Reverend Thomas Malthus's persuasive theory of population growth (first published in 1798) seemed to support that view.

According to Malthus, increased food supply leads to overpopulation: the food supply grows arithmetically (from 1 to 2 to 3 tons per acre, and so on), encouraging people to produce more babies. But populations grow exponentially (from 2 people to 4 to 16 and so on). Therefore, he said, humans are doomed to breed up to the limit of their environment's 'carrying capacity' until epidemic, wars or other disasters rectify the situation. The idea seemed obvious, with its simple mathematical formula (and the London slums did rather seem to be 'teeming'), so Malthus never tested it against real data – which would have shown that people and animals are very good at regulating their own numbers. Later on, people went to extraordinary lengths to find examples of Malthus's model in action. Disney's production team tried to find lemmings demonstrating their famous solution to population pressure for the film *White Wilderness* in 1958, but ended up buying some lemmings, filming them on a rotating snow-covered turntable so that centrifugal force pushed them over the edge, and herding them over a small cliff and into a river for the final shot.[18]

Further to confound the Malthusian model, Martin Jones's eastern Mediterranean examples of environmental collapse date from some thousands of years *after* the appearance of agriculture, and the non-European world is full of examples of agricultures that sustained higher population densities than Europe's without any negative impact on their environments at all – the reverse, in fact. Outside Europe, highly productive, ecologically stable agriculture has been the norm for thousands of years.[19] In many places, peasant agriculture has led to greater biodiversity (Ladakh, which is essentially a high-altitude desert, would have very low biodiversity without its human population, according to Helena Norberg-Hodge[20]).

Malthus would have been right if he had not focused on fertility but had paid attention instead to the role played by debt, which (as Thomas Piketty has shown) necessarily rises in lockstep with rising inequality. Debt becomes an increasingly attractive investment for the wealthy, and increasingly necessary for the poor: a closed feedback loop that, historically, has only ever been ended by a catastrophic economic or political crisis.[21]

Unlike population, debt really does increase exponentially. No matter how hard indebted farmers work to increase production, they can only do so incrementally – but their debts mount in exactly the same, multiplicative way Malthus said human numbers did, thanks to compound interest. As soon as farmers, or producers in general, become indebted to banks, simple arithmetic calls the tune and sets the pace, and producers must do whatever it takes to service the debt – or leave the land, as millions did.

The physicist Frederick Soddy explained it like this in 1926:

> *Debts are subject to the laws of mathematics rather than physics. Unlike wealth which is subject to the laws of thermodynamics, debts do not rot with old age and are not consumed in the process of living... On the contrary, [debts] grow at so much per cent per annum, by the well-known mathematical laws of simple and compound interest... which leads to infinity... a mathematical not a physical quantity.*[22]

And Will Carleton, a poet from rural Michigan, put it this way in the 1890s:

> *We worked through spring and winter – through summer and through to fall*
> *But the mortgage worked the hardest and the steadiest of us all;*
> *It worked on nights and Sundays – it worked each holiday –*
> *It settled down among us, and it never went away.*[23]

Recently, a number of writers (such as the economist Ann Pettifor[24] and the anthropologist David Graeber[25]) have been pointing out that debt was not originally like this: debt and credit were and are essentially something that communities provided for themselves to buffer the ups and downs of life, support new projects, and keep track of who owed what to whom. It was the monopolizing of money by a merchant elite, the right to create debt, and set its terms, that turned debt into the tyrannical, mathematical and even (in the sense implied above by Frederick Soddy) supernatural force that it became. Production, says Ann Pettifor, ceased to be simply to meet needs and wants, but a race 'to make food production *more profitable than debt*'[26] with predictable consequences for people, their communities and, ultimately, the land.

This unwinnable contest was a large factor in the agrarian crisis of medieval and early modern Europe, leading to the exodus of the impoverished – first into the cities (a major trend by the 16th century) and then to the Americas and Australasia in the late 19th and early 20th centuries.[27] Teresa Hayter writes that:

> from the early 19th century to the 1920s, more than 60 million Europeans migrated to America and Australasia, of whom 5.7 million went to Argentina, 5.6 million to Brazil, 6.6 million to Canada, and 36 million to the United States.[28]

Until 2001, this exodus of the European poor was still the largest mass migration in history.

'SHEER HUMAN NUMBERS' RE-EXAMINED

The human story used to be told without its neolithic prelude, so that humanity seemed to emerge from nowhere 5,000 years ago (the dawn of Biblical time), complete with inequality, doomed from the outset to lives of hard labor and counter-productive striving – the Judeo-Christian version of history into which Thomas Malthus's theory seemed to fit quite easily. Yet Malthus never showed any causal link between 'sheer human numbers' and either famine or ecological harm and, when the precise causes of famines, species loss and environmental deterioration finally came to be examined, they failed consistently to implicate population growth. The finger of blame points, instead, at debt, and its essential precondition, inequality.

The Nobel Prize-winning economist Amartya Sen has shown that for every single famine for which records existed, there had in fact been no shortage of food, but a surplus – and a failure to distribute that surplus to the people who needed it. Shop windows might have been full of food, but people simply lacked the money to buy it.[29] Often they were priced out of the market by foreign demand, and that was the situation in all of the great famines that marked the onset of globalization in the 1870s, as it had been in Ireland in the 1840s (where pork and wheat continued to be exported, even past the bodies of the starving).[30] Sen showed that no such famines had ever been recorded in a functioning democracy – often despite extreme climatic and economic challenges.

Cuba could easily have been the scene of famine after the collapse of the Soviet Union in 1989-90 (as North Korea was, after support from China ended at around the same time[31]). In Cuba, it was egalitarian solidarity that enabled the country to switch rapidly from reliance on sugar exports to a high degree of self-sufficiency. As we saw above, this also coincided with environmental regeneration – as it did under a completely different political regime, after inequality fell in Japan three centuries earlier.

'Sheer numbers' were never the issue, but they still have a powerful intuitive appeal – if one believes that a planet is the same kind of thing as a night club. But supposing it were, how many people would fit into it?

In 2008 it was calculated that the world population (then 6.8 billion – but rising to 9 or 10 billion before stabilizing) would fit into a land area the same size as former Yugoslavia, at the same density as Manhattan (in other words, not too dense for comfort, and providing a reasonable amount of open space for recreation). Self-sufficiency would obviously require a lot more space than that – but, not *that* much more: until the mid-1990s, the Chinese city of Shanghai (like other Chinese cities) was self-sufficient in vegetables and grain. Using the population density of Shanghai for that period (2,588 people per square kilometer) the entire world population could (if it wished!) lead an entirely self-sufficient and self-contained, if vegetarian, existence in a land area slightly smaller than the Democratic Republic of the Congo (2.35 million square kilometers: about a fifth of the world's cultivated land[32]), and donate the rest of the planet to wild animals and plants.

This may sound far-fetched, but it is consistent with other studies. For example, a major recent study by Norway's Development Fund (Utviklingsfondet), gathered data from all available sources and found that 25 per cent of the world's food was already being produced within cities before the 2007/8 food-price crisis; and presumably much more than that subsequently.[33] Cities cover less than three per cent of the world's land, excluding Antarctica.

The Norwegian report also shows that modern, intensive agriculture produces far less of the world's food than we think it does – barely 30 per cent, much of which is for animal feed – yet is responsible for 14 per cent of greenhouse-gas emissions (mainly through damage to soils). Fully 50 per cent of world food (including most of the rice we eat in rich countries) comes from peasant agriculture – which has never enjoyed anything like the investment showered on industrial agriculture, and has proved capable of doubling, tripling and even quintupling its yields by organic methods, given modest support.[34]

EHRLICH'S LAST GASP: TECHNOLOGY AND 'EYE-PAT'

In the 1970s, Paul Ehrlich and colleagues offered what many still find a compelling explanation for why, despite all the evidence above, human numbers were destroying the earth: technology. Their argument came packaged in a scientific-looking equation, I=PAT (pronounced 'eye-pat'; environmental *Impact* equals *Population* times *Affluence* times *Technology*). As societies get wealthier, said Ehrlich, they need more cars, washing machines and air conditioners, which means more roads, more raw materials extraction, more fuel consumed, and so on. All of this means that, for the sake of the planet, 'they' cannot aspire to 'our' levels of affluence and technology – unless 'they' reduce their populations drastically.

This may sound hypocritical, but it touches on uncomfortably pervasive, unconscious assumptions many of us have. For example, in Radu Mihaileanu's film *La Source des Femmes* (2011, released in English as *The Source*) the women of a Moroccan village launch a campaign to have the water from their mountain spring piped into the village, saving them hours of carrying heavy loads of water every day. The petty official handling the decision says: 'If we give them their pipeline, the next thing we know, they'll be wanting washing machines and showers'. One flinches slightly at this: most of us use washing machines and showers, know what their environmental cost is, feel guilty about it perhaps... and might like these attractive North African folk to go on living in their beautiful mountain village, untainted by the world of washing machines. An unsettling scenario is evoked, in which villagers and especially the young have to migrate to the city to find work to earn money to supply the new luxuries, the beautiful mountain village becomes depopulated, and its people end up in a grey, dusty, sprawling suburb somewhere... and we feel particularly uncomfortable that this smug, sexist bureaucrat expresses attitudes that we hide even from ourselves.

This vision of inexorable degradation derives from the apparently linear nature of progress: we got where we are along a particular path, and we assume that that is the route everyone will have to take. The engineering scientist Howard Rosenbrock argued against that kind of fatalism.[35] He called it 'the Lushai Hills effect' after the hills in north-east India, which rise from a maze of wooded valleys and ravines. He took a walk there one morning and somehow reached a summit. Looking down, it seemed extraordinary that he had struck on the one possible route to the top. But of course, he hadn't – there were plenty of ways of getting there, all different; and indeed other summits he could have reached.[36]

Most of us have very little say in the decisions that shaped our 'possibility landscapes', yet we must live in them and choose from among whatever options they offer. The greatest fuss is made about the smallest choices (whether to buy Indesit or Electrolux, this phone tariff or that, from this provider or the other), which encourages an illusion of free choice like the one created by close-up magicians, when they ask you to 'pick a card'. More radical choices are either hidden by the sales patter, or under lock and key. Yet they exist.

THE POWER TO CHOOSE A LOW-IMPACT LIFE

I=PAT is promoted as if it were an iron law, decreeing that every improvement must carry an environmental price. But technology is by definition the more and more parsimonious use of natural phenomena: doing more with less. Technological choices should radically reduce impacts – and sometimes they do. I=PAT fails to explain the great differences in impact between countries, shown earlier. Many people in European and Nordic countries, and Japan, enjoy lives just as comfortable as those in the US, but with far lower impact. Even within high-impact countries it is possible to find people who enjoy very high levels of comfort, amenity and health, at very little environmental cost.

In 2012 *The Guardian* interviewed a couple in Lancashire, UK, who have a modest income (only one of them has a paid job, halving their impact at a stroke) but 'feel rich because the quality of life is so rich'. They live in a co-housing project (a co-operatively owned development with shared facilities): 'It's 41 eco houses and we share 11 cars. We cook and shop communally. Because we're sharing everything, we'll have 3 lawnmowers and 6 electric drills instead of 41. We've got shared laundry facilities.'[37]

This very modest, quite unrevolutionary setup is a choice millions of people in Britain would love to be able to have, according to the UK Cohousing network,[38] but only a few hundred can exercise it because the housing industry and regulatory system are not set up to favor small, co-operative ventures. If washing machines have become engines of destruction, it could have more to do with the way housing has been atomized, the machines have been commodified, and daily life restructured to maximize profit by turning communal tasks into private ones – so that people who want clean clothes must buy their own laundries.

Even greater economy and luxury are possible with a bit of political initiative. In nearby Oldham, Lancashire, a small number of families have moved into ultra-low energy 'Passivhaus' housing where they can walk around in t-shirts in the depths of winter yet pay only £20 ($32) a year for heating – compared to the £550 ($880) for gas alone, just for the three summer months, paid by a neighbor in one of the pre-cast concrete council homes they used to live in.[39] These homes only cost an extra £20,000 ($32,000) to build, but Britain only has 150 of them, there is little market incentive to build more, and only market solutions are permitted in the country since public provision of housing was effectively ended in the 1980s. Instead, market incentives favor burning as much fuel as possible: the fuel and power utilities were all privatized and became stock-market listed companies at the same time.

Consequently, most people must live in isolated, expensive, high-impact housing because that's what's available – Britain's rising power inequalities have distorted the 'possibility landscape' in a way that favors the needs of a few extremely wealthy land-owners, construction firms and energy companies over those of everyone else.

The difference between paying £20 a year and more than £2,000 a year, just to keep warm (and the even greater subjective difference between struggling to keep warm and living comfortably and worry-free, in a t-shirt in the depths of a Lancashire winter) shows what a lot of 'give' there is in the I=PAT equation.

We are encouraged to think that reducing impacts is very difficult, and this mindset provides cover for some of the biggest environmental impacts of all: those of the mining and extractive industries, and of globalized freight. If mundane examples like the ones above are kept out of the discussion, it is perfectly possible to be fooled by talk of squeezed profit margins, and of how difficult it is for struggling firms to reduce their environmental impacts as they try to meet the world's 'need' for other 'scarce resources' – including the metals used in electronics. Here, the iron laws turn out to be very generously padded on the supplier's side.

Amrit Wilson tells us that, whereas average return on capital in developed countries is about five per cent, 'in countries like Tanzania, in sectors such as goldmining and oil and gas, even in 1982 it ranged from over 40 per cent to several hundred per cent'; 'less than 0.001 per cent' of turnover is paid in tax. As she says:

> Today Tanzania's workforce is being made available to global capital in new ways for some of the worst kinds of exploitation through

> the establishment of special economic zones (SEZs) and export-processing zones (EPZs). In these zones, which often take over prime agricultural land, wages are low, there are few health and safety regulations, trade unionism and all labor laws are banned, and if experiences in other parts of the world are anything to go by the workers – often predominantly female – face frequent sexual harassment and abuse.[40]

The global spread of SEZs was led by the growth of the global electronics industry in the 1980s. Shenzhen, home of the giant Foxconn plant mentioned at the beginning of this book, was one of the first; in fact SEZs first appeared in the People's Republic of China as a key feature of Deng Xiaoping's reforms in 1978.[41] Having created the conditions for a 'race to the bottom', the phenomenon was turbocharged by the arrival of Information and Communications Technologies (ICTs), which allowed wholesale offshoring of industries such as the garment industry, the shoe industry and call centers.[42] There are now thousands of SEZs around the world. As pressure on raw-material prices has grown, mining has become a 'natural' candidate for SEZ treatment. Profits on the scale mentioned by Wilson are common, together with huge environmental and human impacts.[43]

The environmental costs are deeply ramified, and go hand in hand with the erosion of equality via attacks on workers' security and conditions. In their 2010 documentary *The Forgotten Space*, Allan Sekula and Noël Burch show that containerization of sea freight (which began in New York immediately after the Second World War) was more to do with destroying the power of the dockworkers' and seafarers' trade unions than with commercial considerations. Containerization removes the need for dock labor apart from a few, isolated individuals operating cranes. It allows firms to exert much tighter control over the 'value chain' and even to extend it, exploiting resources and labor much more easily – hence the explosion of 'offshoring', and the amount of energy used in transporting goods.

Containerization is also far less efficient in energy terms than conventional cargo-handling. The bulkier, containerized cargoes require larger ships – which use space extremely inefficiently because containers cannot be stowed deep below decks. The bowels of the ship must therefore be filled with either ballast or heavy, low-value cargoes, which travel almost free – explaining the appearance over recent years of millions of tonnes of heavy stone garden ornaments from China and India in European garden centers. All of this has led to a doubling of the amount of goods carried by sea between 1982 and 2007, so that, by the latter year, shipping carried 90 per cent of world trade. Shipping was by then one of the biggest sources of greenhouse gases on the planet, producing twice as much CO_2 as did airlines, and is predicted to rise by up to 75 per cent within 15 years.[44]

The slowdown in international trade since 2007 has somewhat reduced the growth in container traffic, but any reduction is offset by the growth of oil-tanker traffic: the boom in shale-oil production in Canada and the US has forced traditional oil suppliers such as Venezuela and Nigeria to seek markets in Asia. In 2012, each tonne of oil used in the world had travelled nearly 10 per cent further than it had the year before, making a total of 7.8 trillion tonne/miles for the year, and raising interesting or alarming new security issues. Longer voyages mean more danger from pirates, especially in the seas through which these cargoes must pass, hence 'it is only a matter of time before we see Indian [war]ships in the South Atlantic'.[45]

The same lobby groups that drove through containerization during the 1960s and 1970s also succeeded in having sea transport excluded from the Kyoto Agreement on climate change – making shipping a neatly self-contained object lesson in the environmental impacts of concentrations of power.

1   For example, Alec Nove, *The Economics of Feasible Socialism*, Allen and Unwin, 1983, p 157.
2   Brand, Preston & Boardman, *Travelling in the right direction: lessening our impact on the environment*, Final Research Report to the ESRC 2006: nin.tl/ESRC2006
3   Danny Dorling, personal communication, 28 Sep 2007, citing Worldmapper.org and WWF Living Planet Report data. See also Dorling, *Injustice: why social inequality persists*, Policy Press, April 2010.
4   WWF Living Planet Report, 2008.

5 Brian Pollitt, 'Cuba's health services – crisis, recovery and transformation', *CubaSí* (magazine of Cuba Solidarity, UK), Autumn 2004.
6 Patricia Hynes, 'War and the True Tragedy of the Commons,' *Truthout*, 28 July 2011, nin.tl/PHynes2011
7 P Sahlberg, *Finnish Lessons*, Teachers College Press, 2011. See also: Alan Moore, 'What Makes the Finnish Education System Work?' No Straight Lines, nin.tl/MooreFinland Accessed 16 Feb 2014.
8 Melissa Benn, 'The Question of Private Schools', 9 June 2012, nin.tl/Benn2012
9 James K Boyce, 'Is inequality bad for the environment?' *Equity and the Environment*, 15, 2007, pp 267-288, citing Princen, 'The Shading and Distancing of Commerce', MIT 1997.
10 Ibid.
11 International Panel on Climate Change; for example see graph at: nin.tl/IPCCemissionsgraph Naomi Klein writes that global emissions rose, on average, 1 per cent per year throughout the 1990s, and 'by the 2000s, with "emerging markets" such as China fully integrated into the world economy, emissions had sped up [to] 3.4 per cent a year,.' *This Changes Everything*, Simon & Schuster, 2014.
12 Andrew Revkin, *Global Warming, understanding the Forecast*, Abbeville Press, 1992, p 55.
13 PJ Crutzen & C Schwägerl, *Living in the Anthropocene,* Yale School of Forestry and Environmental Science, 2011. See also Mike Davis, *Living on the Ice Shelf,* Tom Dispatch 26 June 2008.
14 'Forest Razing by Ancient Maya Worsened Droughts, Says Study', Lamont-Doherty Earth Observatory website, 21 Aug 2012.
15 Martin Jones, *Feast: Why Humans Share Food*, Oxford University Press, 2007.
16 Ibid.
17 Jared Diamond, 'The Worst Mistake in the History of the Human Race', *Discover Magazine*, May 1987, nin.tl/Diamond1987.
18 Danny Dorling, *Population 10 Billion*, Constable, 2013, pp 132-133.
19 For example, see FH King, *Farmers of Forty Centuries,* Mrs FH King, 1911; Wayne Roberts, *The No-Nonsense Guide to World Food,* New Internationalist, 2008; and Colin Tudge, *Feeding People Is Easy,* Pari, 2007.
20 Helena Norberg-Hodge, *Ancient Futures*, Sierra Club Books, 1991.
21 Thomas Piketty, *Capital in the Twenty-First Century*, Harvard University Press, 2014, pp 396-420.
22 Frederick Soddy, *Wealth, Virtual Wealth and Debt*, EP Dutton, 1933, quoted by Ann Pettifor, *Just Money*, Commonwealth Publishing, 2014, p 42, commonwealth-publishing.com/?p=201.
23 Will M Carleton, 'The Mortgage', in Ralph Windle, *The Poetry of Business Life: An Anthology*, Berrett-Koehler Publishers, 1994.
24 Pettifor, op cit.
25 David Graeber, *Debt: the first 5,000 years*, Melville House, 2012.
26 Pettifor, op cit, p 43 (original emphasis).
27 The phenomenon of land exhaustion in Europe since the 10th century is described and analysed by (eg) Fernand Braudel, *Civilization and Capitalism,* 1981; Immanuel Wallerstein, *The modern world system,* 1974;

Catharina Lis and Hugo Soly, *Poverty and capitalism in pre-industrial Europe,* 1979.
28 Teresa Hayter, *Open Borders: the case against immigration controls*, Pluto, 2001, p 9, citing estimates by Bob Sutcliffe in *Nacido en Otra Parte*, Bilbao, 1998.
29 Amartya K Sen, *Development as Freedom*, Oxford University Press, 1999.
30 Mike Davis, *Late Victorian Holocausts,* Verso, 2001.
31 *Starved of Rights: Human Rights and the Food Crisis in the Democratic People's Republic of Korea*, Amnesty International, 2004.
32 Joel E Cohen's 1995 estimate of the world's total cultivated area was 13.6 million km^2; see *How Many People Can the Earth Support?* Norton, 1995.
33 *A Viable Food Future, Part 1*, Utviklingsfondet (The Development Fund, Norway), Nov 2011, p 47, nin.tl/viablefuture
34 *Ibid,* p 40.
35 HH Rosenbrock, *Machines with a purpose*, Oxford University Press, 1990.
36 Conversation with Mike Cooley, 2007.
37 Olivia Gordon, 'How much is enough?' *The Guardian Weekend*, 15 Sep 2012.
38 The *UK Cohousing Network* website, cohousing.org.uk Accessed 19 May 2013.
39 John Vidal, 'Actively Cutting Energy Bills in Oldham', *The Guardian*, 1 Nov 2013, nin.tl/Vidalpassivhaus
40 Amrit Wilson, *The Threat of Liberation*, Pluto, 2013, pp 104-105.
41 Hsiao-Hung Pai, *Scattered Sand,* Verso, 2012, pp 165-166.
42 David Holman, Rosemary Batt & Ursula Holtgrewe, Global Call Center Report, 2007, nin.tl/callcenterreport
43 Dionne Bunsha, 'India's Viagra – Special Economic Zones', *New Internationalist*, Sep 2007, nin.tl/Bunsha2007; Aradhna Aggarwal, 'Special Economic Zones', *Economic and Political Weekly* 41, no 43/44, 4 Nov 2006, 4533–36.
44 John Vidal, 'CO_2 output from shipping twice as much as airlines', *The Guardian*, 3 March 2007.
45 Ajay Makan, 'Oil tanker trade soars on back of US boom', *Financial Times*, 15 May 2013.

5

Ever greater impact, ever less benefit: high-tech capital's mysterious lack of growth

Economists have long wondered why more efficient technologies increase human impact instead of reducing it. Computers were supposed to break the trend but seem to have intensified it instead. Yet outside the worlds of electronics, finance and business consultancy, economic growth has ceased. Meanwhile, inequality is the 'elephant in the room', which nobody likes to mention.

Technology evolves through the ever-more parsimonious use of resources: doing more and more with less and less. If that were the whole story, humanity's environmental impact would now be tiny. Yet the complete opposite has happened, again and again, with dismal predictability.

The contradiction was first described by the British economist William Stanley Jevons in 1865, when he studied the way coal consumption rocketed after the replacement of Newcomen steam engines by the several-times-more-efficient engines devised by James Watt and Matthew Boulton, especially through the early 1800s. The 'Jevons Paradox' (often known nowadays as the 'rebound effect') has been observed again and again with every subsequent major technology. A 30-per-cent increase in fuel efficiency of US automobiles since 1980 didn't lead to any decrease in fuel consumption. Mileages increased, and so did vehicle sizes and performance.[1] In 2010, the economist Juliet Schor found that, since 1975:

> Energy expended per dollar of GDP has been cut in half. But rather than falling, energy demand has increased, by roughly 40 per cent. Moreover, demand is rising fastest in those sectors that have had the biggest efficiency gains – transport and residential energy use. Refrigerator efficiency improved by 10 per cent, but the number of refrigerators in use rose 20 per cent. In aviation, fuel consumption per mile fell by more than 40 per cent, but total fuel use grew by 150 per cent because passenger miles rose. Vehicles are a similar story. And with soaring demand, we've had soaring emissions. Carbon dioxide from these two sectors has risen 40 per cent, twice the rate of the larger economy.[2]

It was assumed not long ago that electronics, computers and the internet offered a way out of the Jevons trap. What could be more parsimonious than electronic devices, whose performance seems to increase by almost logarithmic amounts, while their energy needs seem to fall at the same rate? Their components, moreover, get smaller and smaller every year and contain smaller amounts of materials. A well-known electronics writer and futurologist, Ray Kurzweil, once remarked that, if automobile progress had equalled that of electronics over the previous 50 years, 'a car today would cost one hundredth of a cent and go faster than the speed of light'.[3] The promise of a 'weightless' or 'frictionless economy' became an unchallenged orthodoxy. Respected economists urged large-scale investment in ICTs (Information and Communications Technologies) by businesses and governments.

Alas, Jevons rules here as well. As we will see below, the drive for smaller and smaller, faster and faster chips has been a largely self-defeating, runaway phenomenon that delivers only modest improvements in performance for the people who use them while imposing constant replacement costs on everyone who wants to 'stay in the game'. Power-saving technologies (like lithium batteries, high-density chips, and tantalum capacitors) also consume greater amounts of energy in manufacture, as do the new techniques and chemistries required in the quest for smallness and speed (see Chapters 8 and 9). Computers have played a major part in accelerating human impacts.

'KEEP YOUR NERVE' OR 'TOUGH IT OUT'

Two not-very-satisfying answers have been offered to this problem, both of which boil down to doing nothing about it, but for a choice of two reasons. The first is that we should live with the problem and trust that it will sort itself out. The second is that we should 'carry on regardless' on the principle that one cannot make an omelet without breaking eggs and, if we lose a few hundred species and change the climate, so what? Change is inevitable! We have to be hard-headed realists and commit all our energies to the fight, as 'nature red in tooth and claw' intended.

Juliet Schor has described how the latter approach has been justified in some quite detailed arguments by various mainstream economists. William Nordhaus, for example, argues essentially that since future generations will certainly be wealthier than we are, and will have much more efficient and powerful technologies, it is much better to leave any clean-up operations to them, and concentrate on producing the economic growth that will ensure their future wealth.[4] 'Biodiversity offsetting' is a subtler version of the same logic: ancient woodlands, for example, are notionally replaced by larger numbers of new trees, planted in more convenient locations.[5]

The former 'keep our nerve and it'll sort itself out' tendency has clung tenaciously to a theory published by the Ukrainian-American economist Simon Kuznets in 1953. Kuznets, a passionate anti-communist, recognized that inequality posed a serious moral threat to capitalism, but he thought he had found, by studying a number of economies over time, that although wealth inequality grew during the early stages of industrialization, it decreased thereafter; rising inequality was simply the birth pangs of a new age in which inequality would fall.

At least, as the French economist Thomas Piketty explains, Kuznets *hoped* that inequality would fall; he originally regarded his theory as 'perhaps 5 per cent empirical information and 95 per cent speculation, some of it possibly tainted by wishful thinking', but it was so enthusiastically received that he shed his doubts and the 'Kuznets Curve' soon gained the status of a natural law.[6] It did indeed seem somewhat plausible in the US in Kuznets' heyday (a period of lower inequality) and the theory earned him a Nobel Prize in 1971, but then, growing inequality returned to the US and its most closely linked economies.

The Kuznets Curve remains a powerfully attractive idea, however, with strong, intuitive appeal. There are many situations where 'keep calm and carry on' is a good rule and, latterly, it has been used to argue that economic growth will be good for the environment in the end. Rich economies seemed not long ago to have reduced their consumption of messier raw materials like iron ore and coal.[7] Yet this, too, turned out to be an illusion: 'offshoring' of heavy industry had simply shifted consumption overseas. Jevons remains in control.

Gavin Bridge, a materials scientist from Manchester, points out that while emissions of the gas that causes acid rain, sulphur dioxide, and the particulates (from vehicle and generator-exhausts) that cause smog, decreased by 56 per cent in Europe and 37 per cent in the US between 1980 and 2000, Asia's sulphur-dioxide emissions increased by 250 per cent over the same period: the time when the US and Europe 'offshored' so much of their manufacturing industry.

In 2011 a British climate researcher, Chris Goodall, thought he had spotted a 'Kuznets transition' in Britain: consumption of various key inputs seemed to have fallen slightly in the years before the crash of 2008. He particularly pointed to a reduction in the UK's consumption of paper, which he attributed to the adoption of 'more sustainable' electronic technologies. Similar claims were made in 2014 by a British politician, Chris Huhne, in an article entitled 'Don't Fear Growth – It's No Longer the Enemy of the Planet'.[8]

But, as Keele University's Andrew Dobson promptly pointed out, the dream of 'green growth' is a complete illusion.[9] It depends on ignoring costs that were 'offshored' when manufacturing was moved overseas, and the full cost of 'clean' technologies. Huhne used LED lights as an example (Light-Emitting Diodes). These need impressively little energy when you are using them, but manufacturing them takes large amounts of energy and materials: they are microchips, made by the same high-energy processes as computer chips (described in more detail below, in Chapter 9). They are a very good example of how energy consumption and environmental impacts have to a large extent merely been displaced. The idea of 'green growth' could be just another reincarnation of the old belief, like the belief in perpetual motion machines, that can have one's cake and eat it.

Kuznets' theory belongs with the *faux*-Darwinian mindset mentioned earlier, which pervades market thinking: all is well, natural forces are at work behind the scenes, and everything will work out fine if we allow things to take their course and don't rock the boat. This has become a very strong theme where the more worrying impacts of new technology are concerned. The philosopher Slavoj Zizek calls it 'millenarian apocalyptism':

> none of this suffering and mayhem matters – we are all moving on anyway, leaving it all behind. From the vantage point of our post-human future, all this will be seen as simply the detritus of the larval stage from which we have moved on.[10]

Which leaves the way clear for the sort of hard-core social Darwinism voiced – initially only in private but then more openly – by such figures as the economist Larry Summers who, in 1992, as Chief Economist to the World Bank, suggested in an internal memo that the Bank should encourage the export of dirty industries to poor countries:

> A given amount of health-impairing pollution should be done in the country with the lowest cost... I think the economic logic behind dumping a load of toxic waste to the lowest-wage country is impeccable and we should face up to that.[11]

Summers may have meant this as some sort of thought experiment but, when it was leaked to *The Economist*, it was taken quite seriously. While expressing some distaste for the memo's 'crass' style, *The Economist* opined that 'on the economics his points are hard to answer'. Summers was subsequently made Secretary to the US Treasury and then President of Harvard. Apparently, 'let them eat shit' was far less damaging to one's career at the dawn of the 21st century, than 'let them eat cake' was at the end of the 18th.

It seems very hard, however, to imagine a world containing iPhones that does not take that kind of attitude to other people's health. It is part of the 'no pain, no gain' mentality that was ground into the Western mind centuries ago. But what if the pain no longer yields any gain?

WHY COMPUTERS HAVE GROWN NOTHING BUT THEMSELVES

Computerization was supposed to have been a 'third industrial revolution', the previous two having been driven by steam, from the late 1700s, and by electricity, from the 1890s: periods of furious economic growth that changed the quality of people's lives in radical ways. Yet, despite the spectacular growth in the numbers of computers, especially from the 1980s onwards, conventional economic growth has been conspicuously weak or absent from the wealthy countries' economies over that period. Two economists, Robert Solow (winner of the 1987 Nobel economics prize for his pioneering work on the analysis of economic growth) and Robert Gordon have repeatedly found that the only substantial economic growth since the emergence of the computer industries has been in the computer industries themselves.

Admittedly, economic growth is not a good measure of a population's well-being. It reflects the amount of financial transactions in an economy, which may merely indicate that life has become more expensive, without increasing anyone's material well-being (for example, people may have to travel further to work, or there is more crime, leading to higher insurance premiums and more sales of burglar alarms, and so on). All the same, and especially if we use growth as a measure, you would expect the massive computerization of the past 30 years to have made a positive difference to the figures.

Instead, since the 1980s, economists of all persuasions have been puzzled by the sluggish growth of the richer economies. The literature on this puzzle almost forms a discipline in its own right.[12]

In 1987, Robert Solow stated that 'we can see the computer age everywhere but in the productivity statistics'. Henceforth, this was known as 'Solow's paradox'.[13]

Robert Gordon has spent the past 30 years analyzing the impacts of different waves of innovation on economic growth, unemployment and the material conditions of people's lives, and he has repeatedly found an extraordinary failure of ICT industries to have much positive effect.

Gordon found that in the mid/late 1990s there was an impressive-looking spurt of growth in the US economy which might have been attributed to ICTs, but, on closer inspection, he found that it affected just 12 per cent of economic output: the sectors dominated by computer equipment and peripherals, and the kinds of consumer durables that had been given a new lease of life by electronics. In other words, the ICT-related boom, such as it was, was confined largely to sales of ICTs, which 'raises the question of how far the New Economy actually reaches into the remaining 88 per cent of economic activity'.[14]

Turning to investment rates, these too seemed to have experienced an upswing – yet, again, much of the investment was in computers and computer peripherals, which depreciate rapidly. (And why *do* they depreciate so rapidly? There is no Law of Nature decreeing that they should!). In 2012 Gordon wrote:

> *Invention since 2000 has centered on entertainment and communication devices that are smaller, smarter and more capable, but do not fundamentally change labor productivity or the standard of living in the way that electric light, motor cars, or indoor plumbing changed it.*[15]

The computer-assisted boom in financial services did not deceive Gordon: 'If computers truly raised the output of these intermediate industries in unmeasured ways, then the benefits should show up in the output of final goods... Yet this spillover from intermediate to final goods industries is just what cannot be found in the official data.'

The point was made when the financial boom collapsed in 2008, revealing a shrunken underlying economy. The financial boom's real effect turned out, in the end, to have been to redistribute wealth to the richest one per cent, whose share of US income had risen from 12 to 25 per cent, and share of US wealth from 33 to 40 per cent, in the quarter century to 2011.

Similar effects were found elsewhere, especially in more unequal countries, like the UK. A 2011 study by Manchester University Centre for Research on Socio-Cultural Change (CRESC) found that the financial sector's contribution to the economy had been largely illusory.[16] During the boom years of 2002-8 it contributed only £193 billion in tax (compared to £378 billion by Britain's emaciated manufacturing sector) and employed around one million (mostly low-paid office and call-center workers) compared with two million in manufacturing. In 2007, the height of the boom, around 40 per cent of its lending went straight into property and a further 25 per cent went to financial intermediaries (who probably also put the lion's share into property). Of the remaining 35 per cent, very little found its way into productive enterprise.

A feature of the computerization boom not mentioned by these critical economists (but which looms very large among computer professionals) is the enormous and largely unproductive cost imposed on computer users by the need for continual changes to software. These are partly the consequence of the sub-optimal route computer development has taken (as we will see in coming chapters) and range from the regular 'tax' levied on users via the need to upgrade their systems (mainly so that they can carry on doing what they did previously), to security scares (stemming from the inherent vulnerability of highly hierarchical systems – as US national security whistleblower Edward Snowden pointed out in 2013) to multi-billion dollar write-offs when software projects fail.

Computer project failures are rarely reported when they happen to commercial firms but the costs filter through onto the bottom lines nonetheless, and thence into the GDP figures, where they contribute to the perception of economic growth. They *are* reported, however, when they happen in the public sector, where they give the impression of public-sector profligacy or incompetence. In 1994 the US Federal Aviation Administration wrote off $2.6 billion on a failed air-traffic-control software project, and the IEEE article in which that is reported carries a list of about 50 similar disasters from previous years, warning that 'the problem only gets worse as IT grows ubiquitous'.[17] In 2010 the British National Health Service had to write off £10 billion for an undelivered system for managing patients' records[18], and one could fill the next page or two with similar catastrophes and similar price tags.

The point in emphasizing the scale of these public-sector losses is that they are only a tiny sliver of the overall problem: as Thomas Piketty pointed out in *Capital in the Twenty-First Century*, the public sector is now a far tinier fraction of the total economy than even many economists realize.[19] To appreciate the full social cost of this hidden hemorrhage of wealth we would need to multiply the public-sector losses 50- or even 100-fold (and perhaps more, as public services are privatized or part-privatized by outsourcing to private firms, whose internal affairs are hidden behind commercial confidentiality clauses). Ideally, we should then add the consequential costs ranging from lost production to the mounting cost of risk-management services and insurance.

In 2009 the Standish research group reported that, across the entire US economy, only 32 per cent of computer projects were being delivered on time and on budget, while 44 per cent were running 'late, over budget, and/or with less than the required features and functions and 24 per cent [were] cancelled prior to completion or delivered and never used.'[20]

The overwhelming majority of these costs are inflicted on large, centralized organizations by other large, centralized organizations, in a world increasingly dominated by large, centralized organizations. We can see this from the rise of management consultancy as a sector of the economy, and of firms that provide it, such as McKinseys, and firms that previously only provided accountancy services such as PricewaterhouseCoopers, DeLoitte, KPMG, Ernst and Young, and Accenture (known as Arthur Andersen Consulting before its implication in the Enron scandal of 2001). They are the work of a very small elite of wealthy, powerful men (and a few women) who deal only with other wealthy, powerful men and women, making huge promises to each other on complicated matters that they cannot claim to understand.

INEQUALITY: THE ELEPHANT IN THE ROOM

The problem is not that such consultants are stupid, or necessarily crooked, but that the elevated positions they occupy mean they cannot afford to appear anything less than totally in command of the situation and confident of its outcome. This is a general, institutional problem of inegalitarian economies, and an extremely intractable one – as economist and psychologist Daniel Kahneman showed (with particular reference to the 2007/8 financial crisis) in his book *Thinking, Fast and Slow*.[21] The human mind's obdurate tendency to overconfidence becomes a major liability for humanity when the decision-making is delegated to tiny numbers of highly privileged minds.

It is not just a problem of elites. Everybody in an unequal society is precarious, so nobody can afford to be entirely honest. Those who have a lot, have a lot to lose; those who have little can easily lose everything. Before any effort can go into producing things, or caring for others, or just enjoying oneself, one must make sure one's economic position is secure, by whatever means that entails. Resources and ideas must be deployed primarily to defend positions of power or safety. This is an iron law that even the most philanthropic entrepreneur must obey, or perish. To compound the problem, the climate of mutual watchfulness makes it impolite to discuss the inequality that animates the system. Making any kind of plan becomes like trying to discuss travel options between Italy and France, without mentioning the Alps.

The medical evidence that led to the discovery that inequality *per se* makes us ill, derives precisely from the strain of coping with these social gradients, which our physical bodies must deal with despite powerful social conventions prohibiting their discussion. It's as if, instead of helpful gradient warnings, the road signs here all said:

> 'Pay no attention to this cliff. Maximum penalty, death!'

Despite the attention that has been given to books like *The Spirit Level*,[22] we are a long way from having an open discussion of inequality, unless the conversation is kept to a very general, almost abstract level. If anyone doubts this, they should try striking up a conversation with their employer about his salary, pension, investments, family wealth, and where he sends his children to school and why... better still, invite yourself around to his house to chat about these things with a colleague or two. (This may not seem such a massively challenging prospect to someone who has lived in Denmark or Sweden.)

Conversely, the powerful are able to use their lofty perches to scrutinize the personal affairs of those below them in more and yet more intimate detail. Then, because lofty perches are not legitimate subjects of discussion, they are free to play the equality card when it suits them, demanding equal rights, for example, to use their money to support the causes they believe in. (This was the basis of the 2010 'Citizens United' case, in which wealthy rightwingers successfully argued that restrictions on private political donations discriminated against them.)

The barriers to discussion are endemic, structural, and become stronger as inequality increases. Rapid increases in private wealth in India, for example, have been accompanied by deterioration in tax data[23] – perhaps as a result of increasingly effective lobbying against what the super-rich see as intrusive enquiry. In an earlier period of intensifying inequality, the late 1890s, proposals by the French government to gather data on personal wealth were denounced as 'inquisitorial'[24], and similar language has become familiar again among the neoliberal elite in the US and UK.

Commercial confidentiality and property rights of more and more kinds are increasingly strongly enforced (as discussed later, in Chapter 8) and backed by the powers of nation states and international treaties. When state protection of sensitive information is insufficient, firms and wealthy individuals can move their wealth and affairs to tax havens, which no nation state so far has dared seriously to challenge – with the interesting consequence that up to 10 per cent of the world's wealth cannot now be accounted for. According to official statistics, rich countries are in 'asset deficit': fewer of their assets are owned by residents than by foreigners. Logically, this should mean that poorer countries are in asset surplus, yet they, too, are in deficit. Piketty comments: 'It seems, in other words, that Earth must be owned by Mars'.[25]

1 John Bellamy Foster, Brett Clark & Richard York, 'Capitalism and the Curse of Energy Efficiency', *Monthly Review* 62, 6, Nov 2010.
2 J Schor, *Plenitude,* Scribe Publications, 2010.
3 Giles Slade, *Made to Break,* Harvard University Press, 2006, p 197.
4 Schor, op cit, pp 78-9.
5 George Monbiot, 'Reframing the Planet', 22 April 2014, nin.tl/MonbiotApr2014
6 Thomas Piketty, *Capital in the Twenty-First Century*, Harvard University Press, 2014, pp 11-26 and notes p 581.
7 T Jackson, *Prosperity without Growth,* Earthscan, 2009.
8 Chris Huhne, 'Don't Fear Growth – It's No Longer the Enemy of the Planet,' *The Guardian*, 24 Aug 2014, nin.tl/Huhne2014
9 Andrew Dobson, 'Debunking Chris Huhne's Paean to Growth', *The Guardian*, 28 Aug 2014 nin.tl/Dobson2014.
10 Slavoj Zizek, *First as Tragedy, Then As Farce,* Verso, 2009, p 94.
11 SA Marglin, *The Dismal Science,* Harvard University Press, 2008, p 37.
12 For example: Steven J Landefeld & Barbara M Fraumeni, *Measuring the New Economy in the United States,* 2001, nin.tl/1RjBotF; Martin N Bailey & Robert L Lawrence, 'Do We Have a New Economy?' paper presented at the annual meeting of the Allied Social Science Association, 5 Jan 2001, and in *American Economic Review* (May 2001); and Karl Whelan, 'Computers, Obsolescence, and Productivity,' Federal Reserve Board, Feb 2000.
13 Robert J Gordon, 'Does the "New Economy" Measure up to the Great Inventions of the Past?' *Journal of Economic Perspectives* 14, no 4 (Fall 2000): 49–74.
14 Ibid.
15 Robert J Gordon, 'Is US Economic Growth Over?', NBER Working Paper, Aug 2012.
16 Ewald Engelen, Julie Froud, Sukhdev Johal, Adam Leaver & Karel Williams, *After the Great Complacence,* Oxford University Press, 2011.
17 Robert N Charette, 'Why Software Fails', *IEEE Spectrum*, 2 Sep 2005, nin.tl/softwarefails
18 Rajeev Syal, 'Abandoned NHS IT System Has Cost £10bn so Far', *The Guardian*, 18 Sep 2013, nin.tl/NHSITsystem
19 Piketty, op cit, p 48.
20 'Project Failures Rise – Study Shows', nin.tl/projectfailures Accessed 12 Sep 2014.
21 Daniel Kahneman, *Thinking Fast and Slow*, Farrar, Strauss & Giroux, 2011.
22 Richard Wilkinson and Kate Pickett, *The Spirit Level*, Penguin, 2009.
23 Piketty, op cit, p 329.
24 Ibid, p 636, note 23.
25 Ibid, p 465.

6

The invisible foot: why inequality increases impact

Inequality and environmental impact go hand in hand, but what is the causal mechanism? Is it human nature always to want more? Is it a matter of 'keeping up with the Joneses'? This chapter introduces the idea that our excess consumption is largely forced on us by 'positional' forces that are beyond our individual control, unleashed by policies that prioritize competition over community.

Whether or not the market has an 'invisible hand', as the 18th-century economist Adam Smith maintained, today's market economies certainly contain an 'invisible foot': inequality, which bears down on all people of all classes, forcing them into choices that often seem irrational.

Ignoring inequality's role leaves an explanatory void, which is then filled by psychological and moralistic judgements. Environmental activists often argue that the climate crisis is caused by a failure of people in general to 'wake up' to the problem, and that anyone, no matter how poor, can live in a sustainable manner if they really put their minds to it.[1] Or people argue that human nature itself is the problem (so draconian solutions are needed, or we are simply doomed).

The Victorian economist William Stanley Jevons (mentioned in the last chapter) thought that people had an inexorable tendency to consume more when more was available. In 1880, he wrote:

There is no end nor limit to the number of various things which a rich man will like to have, if he can get them. He who has got one good house begins to wish for another: he likes to have one house

> in town, another in the country. Some dukes and other very rich people have four, five, or more houses. From these observations we learn that there can never be, among civilized nations, so much wealth, that people would cease to wish for any more. However much we manage to produce, there are still many other things which we want to acquire.[2]

Few mainstream economists challenged this view. But we now know that rich people's acquisitiveness is not as strong everywhere, at all times. For example, in Britain and the US in the 1950s and 1960s, top executives were content with much more modest salaries and lifestyles than those they sought, with increasing success, in the 1980s and beyond. Their mushrooming salaries had nothing to do with radically increased productivity – the reverse if anything. Instead, as Thomas Piketty has shown, their salaries soared purely because of a radical transformation in the tax structure – from the near-confiscatory rates that had been introduced during the Second World War (98 per cent for highest earners in the UK in 1941-52 and 1974-78) to something approaching a flat tax rate of 30-40 per cent, after the rightwing shift of the 1980s.[3]

Before the 1970s, big increases were fairly pointless, because most of the extra income would go in tax. Afterwards, there was lots for top executives to go for, and they went for it (no doubt reassuring themselves that if they didn't, others would). Their 'human nature' changed with the tax regime, and with the intellectual climate that developed around it. A similar situation prevailed when Jevons was writing: inequality within advanced European economies was riding at extremely high levels (similar to the global level today) and there were almost no constraints on individual wealth accumulation (as Piketty points out, this is also the case nowadays at the global level).

Another explanation blames humanity's weakness for 'keeping up with the Joneses'. Adam Smith described an obsession with 'frills and trinkets'; anthropologists speak of 'status goods': things you buy to feel good about yourself, and avoid feelings of shame. In his 1899 book *The Theory of the Leisure Class* the

The invisible foot

US economist Thorstein Veblen coined the term 'emulative consumption', describing how, in an unequal world, 'everyday life is an unremitting demonstration of the ability to pay'. In 2008 *Le Monde*'s environmental editor, Hervé Kempf, applied Veblen's analysis to the environmental crisis in his book *How the Rich are Destroying the Earth*. Kempf showed that the rich not only spur each other on in their amazing feats of overconsumption, but also transform consumption all the way down the social pecking-order. He also draws links between rising inequality and state violence and the erosion of democratic rights via interlinked processes: the concentration of media ownership in the hands of a rich elite, and the globalization, since the fall of the Soviet Union, of a culture that promotes individualistic aggression and consumption. His analysis is fact-filled and persuasive (the original French edition from 2007 sold over a million copies). His hopes are pinned on a reborn European Left 'unifying the causes of inequality and ecology'.

Emulation is a powerful force, but it is nothing like the whole story. This chapter and the next one seek to 'de-psychologize' the link between economic activity and impact, and show how inequality, in and of itself, makes it difficult or even impossible for people to avoid environmentally and socially destructive choices, irrespective of any moral or intellectual deficits they may or may not have. We find a continuum of compulsion at work, ranging from the family that buys clothes it suspects were made in a sweatshop (because that's all they can afford, and alternatives are extremely expensive), to the landless people who cut down trees in the Amazonian forests, even though they know exactly what the consequences will be (because, as the Brazilian activist Chico Mendes pointed out, if they didn't do it, their families would starve[4]), to the chief executive of a British manufacturing company who transfers production to a Special Economic Zone in a poorer country, even though it violates his instincts (because if he didn't his firm would be out of business).

To be sure, personal morality and courage can overcome all of these things – but only to the same extent that strength of character will get you up Mount Everest.

TECHNOLOGY PLUS INEQUALITY EQUALS MELTDOWN

Much if not all so-called emulative consumption is to a greater or lesser extent enforced – for example because car ownership has become so prevalent that normal life is no longer possible for someone who holds out against it. This kind of coercion is an integral feature of unequal societies because they are unconducive to the sort of collective provision (for example, of public transport) that would make the decision to buy a car or not less important. Expectations are important. The feeling that society is *about* to become less equal forces people to consider individual solutions that work against the common good. The ethos can change quite suddenly from 'all hands to the pump' to 'every man for himself'.

The economist Fred Hirsch explored this kind of economic activity in a 1977 book called *Social Limits to Growth*. Like Richard Wilkinson (whose discoveries about health and inequality began at about the same time), Hirsch had noticed a flaw in the story of rising post-War prosperity. Incomes had risen immediately after the War and, at first, so had living standards. But from the early 1960s, while incomes continued to rise, living standards failed to go up at the same rate. More and more of the extra income was being spent on what he called 'positional goods': ones whose value is reduced, or which cease to be luxuries and become necessities, when enough other people have them as well.

Positional goods such as cars are, of course, very desirable as well as useful. But it is an overstatement to say that people only buy and use them because they have 'a love affair with the car'. A recognizable example might be someone who would rather use public transportation, and is ideologically committed to it, but who buys a car. She does so not because she is confused or hypocritical, or has fallen in love with cars, but because it has become impossible to do a week's shopping, be an adequate parent and do a full-time job otherwise (her work, and the nearest adequate supermarkets, are now several bus-rides away, having relocated to more profitable out-of-town locations on the grounds that 'everyone has a car these days'; taking a family on holiday or just for a day out on public transportation has become a major challenge for the same reason).

There's a similar mixture of motivations in computer and mobile-phone ownership – with the element of coercion perhaps more salient. Once enough other people own one, it becomes increasingly hard to do without one. More and more firms and public services assume that you have access to the internet. Jobs are advertised and often must be applied for online, so to get a job you often need a computer or access to one.

The element of coercion is most extreme at the bottom of the economic pile. A 2012 report by the mobile-phone operator O2 found that more than a fifth of young people in Britain would rather go without food than without their phone, so vital had it become to preserving quality of life.[5] In the same year, a World Bank study said that among those earning less than $2.50 a day (known in World Bank jargon as 'the base of the pyramid', or 'BOP') it is not unusual to spend a quarter of household income on mobile-phone charges. The study found that in Kenya 'at least 20 per cent of respondents felt it was necessary to make real sacrifices to recharge their mobile credit. In the majority of cases (more than 80 per cent), that meant buying less food, at least once a week.'[6]

'POSITIONALITY' AND 'HUMAN NATURE'

Positionality is a valuable concept for anyone who is interested in making society more fair, because it highlights the fact that the system is not composed entirely of people who believe in it and want it, but perhaps very largely of people who have been dragooned into it and would love to be free of it. This may even be true of some members of the rich elite. Positionality, said Hirsch, turns even a wealthy economy into a 'frustration machine'.

Hirsch died in mid-career, not long after his book was published. The economist and writer Robert Frank revived the concept in his 2007 book *Falling Behind: How Rising Inequality Harms the Middle Class*.[7] Frank's phrase for this kind of consumption is 'smart for one, dumb for all' – but it's important to recognize that people do not necessarily embark on this kind of consumption because they are dumb, but because they have to.

Positionality affects our behavior deeply, insidiously and inexorably. It makes human societies destructive irrespective of what kind of people we think we are, by 'tilting the landscape' so that it becomes extremely difficult to behave in a virtuous way without heroic personal effort. Positional behavior achieves only marginal or no improvement in what one gets out of life, yet it requires a greater and greater expenditure of energy and resources to maintain it. As we will see, this expenditure escalates as more powerful technologies are brought into play. Furthermore, once one is caught up in positional competition there is no way an individual can easily escape from it, unless society as a whole decides to call a halt to the proceedings.

To borrow a signoff that was current in the early days of the World Wide Web: 'when the avalanche has started, it's too late for the pebbles to vote'.[8]

Hirsch thought the positional principle applied just to a subset of goods that were intrinsically scarce, but (to take one of his examples) the scarcity of beautiful places in which to live is by no means self-evident. Cities are quite capable of being beautiful, for example. What is more, the reason why so many of them are so ugly is a positional one: they are built by construction firms whose design options are constrained by the practices of their rivals. Once one firm finds a cheaper way of building, others must follow suit, even if the result is a bit less attractive. Should one of them decide to buck the trend and build the kinds of houses people prefer (generally, ones that contain more of the kind of detail and variety produced by human labor[9]) they would either have to cater exclusively for a very wealthy elite, or go bust. It is certainly also true that nasty, grasping developers who despise popular taste play their part – but even they are a positional phenomenon. When a firm's survival may depend on decisions that go against the common good, having a sociopath at the helm can offer significant competitive advantage, at least in the short term; by the same token, a non-psychopathic chief executive might be a risky choice.

Positionality describes a very wide range of familiar situations where one person's gain is another's loss: 'diminishing returns' situations; ones where people find they have to run to stand still, push to the front or be left behind; and 'tragedies of the commons', in which a common resource goes up for grabs and no individual can change things, or avoid losing out, by refusing to join in the plunder. This kind of activity is often called 'competitive behavior' and blamed on human nature. But, often, it goes *against* our nature. We feel ashamed of these actions and inactions, if we accept personal responsibility for them (as we are constantly encouraged to do), and this is politically disabling.

People who occupy different parts of the wealth spectrum are likely to believe wildly different things, exacerbating the situation. The beliefs found in the highest echelons can be wildest of all – and these people are able to deploy enormous resources (including those of the state) to defend their notions of reality, and exclude, stigmatize and undermine rival ideas. Piketty often refers to the importance of the 'justificatory systems' that develop around inequality. These can become more and more bizarre. At the peak of European inequality in the late 19th century, the popular and influential French economist Paul Leroy-Beaulieu maintained that his country was well on the way to achieving the equality proclaimed in the French Revolution, and that introducing even a mildly progressive tax would stop this desirable development in its tracks.[10] Similarly, in 2015, it seemed a complete mystery to many people that economic policy was still based on an alleged need to cut the incomes of the poor and boost those of the rich still further, supposedly in the interests of general prosperity.

TRAFFIC WAVES AND WHY FASTER IS SLOWER

The concept of positional consumption makes it possible to see the problem of human impact in more broadly scientific terms, as one of many phenomena resulting from energy being poured into a system that cannot absorb it adequately – for example, the 'bow-wave effect' known to sailors and naval architects, which negates efforts to drive a ship faster. As the vessel's speed increases, the wave at the bow gets bigger, requiring a disproportionate increase in energy to overcome its resistance. Beyond a certain point ('hull-speed'), there is no point driving a vessel harder because its bow-wave weighs as much as the ship itself. You can open the throttle as wide as you like but all you will do is generate heat, carbon dioxide, noise and vibration, and deafen the fish.

'Traffic waves' are another example, which most of us now experience from time to time in the form of hold-ups that have no apparent cause. These are like the standing waves that form in rivers – easily appreciated from a traffic helicopter or on a test-track, but completely mysterious to those affected. They explain why it is impossible for a transport system based on private vehicles to achieve anywhere near the efficiency of a centrally organized one (like a railway or, even more, the internet, with its huge throughput of data packages over cables that were once thought incapable of handling so much traffic). They also provide a key to other major problems of liberalized economies, such as boom-bust cycles and speculative bubbles – and, of course, their environmental impacts.

Traffic waves might be blamed on bad or aggressive drivers, whose behavior disrupts the flow of traffic, but an experiment carried out in Japan in 2008 showed that even very good drivers, earnestly trying to co-ordinate their efforts, are prey to the phenomenon. The least fluctuation in the velocity of one vehicle sets up a chain reaction that amplifies the original perturbation until everyone is at a standstill, irrespective of the drivers' personal qualities. The 'human nature' explanation proves, again, to be a red herring. The experiment involved stationing 22 identical cars at equal intervals on a circular test-track, and instructing their drivers to cruise steadily at 30 kilometers per hour (kph). No matter how hard the drivers tried to maintain distance and speed, the cars always ended up in a slower-moving cluster, which always came to a momentary standstill – which moved backwards, like a wave, against the direction of travel at about 20 kph – the same speed observed on motorways.[11]

German sociologist and physicist Dirk Helbing has built computer models of traffic waves,[12] and argues that they are part of a much larger category of phenomena: the 'self-driven, many-particle systems' that include fluids, flocks and herds of animals, and financial markets, with their analogous tendencies to boom and slump, even in the absence of an actual shortage or glut. Helbing has also studied pedestrian behavior, which contrasts with that of vehicles in its much stronger tendency to self-organization – people spontaneously form themselves into streams in each direction on crowded walkways, so that hold-ups are avoided, except in very extreme and unusual disaster situations.

A speeded-up system composed of self-directed elements consumes more and more energy for less and less benefit (a 'faster is slower' effect) and can even lock itself up solid (the phenomenon known as 'freezing by heating': very apt considering that it is happening in an economy that has more or less frozen solid, in the sense of meaningful economic activity, while heating its surroundings to a perilous degree).

During the past century, the *positional* nature of the automobile has provided the leitmotif of a very general positionalizing trend, which has changed the nature and fabric of life. First, the process of suburbanization and urban sprawl, as car ownership increases; then the rearrangement of the economy's infrastructure with its motorway networks, container ports, and so on, while the liberalization of investment through privatization, marketization and globalization of trade rules has 'freed up' other aspects of life to behave in the same way. Computers and electronics have made it possible to intensify and accelerate these processes, and extend them into every area that can be monetized (see Chapter 7).

COMPUTERS AND THE POSITIONAL ECONOMY: OBSOLESCENCE GONE MAD

Obsolescence is like suburbanization in that your environment changes as the result of the actions of others – and obsolescence has burgeoned since the arrival of microelectronic devices, and their incorporation into products.

We associate microprocessors and microchips with computers, yet these contain a mere two per cent of all the chips that are produced. The rest go into appliances of every kind and size.[13] By

2000, industry journals reported that '20 per cent of the cost of a luxury car is now in the electronics'[14] and that 'The Volvo S70 has not one, but two CAN [Controller Area Network] buses running through it, connecting the microprocessors in the mirrors with those in the doors with those in the transmission. The mirrors talk to the transmission so that they can tilt down and inwards when you put the car in reverse. The radio talks with the antilock brakes so that the volume can go up and down with road speed (the ABS has the most accurate speed information)'.[15]

Efficiency, reliability and comfort are improved to varying degrees, but you can't repair a computerized car or washing machine as easily as a pre-electronics model. Very likely it would not be cost-effective to repair it anyway, because a whole sub-assembly would be needed, costing more than the machine is worth. Many car drivers would like to be able to buy vehicles without electronics (and certain vehicles, like pre-electronics versions of the Toyota Hilux, continue to command premium re-sale prices). But as soon as most manufacturers adopt electronic subsystems, all must do so, for a plethora of reasons: not to do so would mean retaining old and increasingly hard-to-replace productive capacity and skills, forgoing all the highly promoted sales advantages of the new technology, and allowing lucrative after-sale income (from servicing and repairs) to flow into the pockets of small garages, or be avoided by skilled amateurs, instead of being retained by the company via the car-specific computer diagnostic systems that come as part of the electronics deal. This in turn has a knock-on effect on the wider society, because small garages and repair shops are cut out of the food-chain, and disappear because there is less and less in a car that they can repair.

When it comes to personal computing devices, obsolescence becomes extreme. The normal life of a personal computer is now less than four years, and two years for a smartphone – and this is not because people are slaves to novelty. As soon as any new development becomes widely adopted – new services like social media, or a new form of connectivity or data storage – a whole generation of machines is made obsolete. No manufacturer can afford to continue to fit machines with older storage or connectivity options because that would destroy their slim profit margins.

The manufacturers, while benefiting from being able to sell new versions of the same products to the same customers every couple of years, also have obsolescence forced on them by the constant obsolescence of their basic components, their microchips, as chip manufacturers battle to increase speeds and densities (I will explain how this works in Chapter 8). Chip manufacturers in turn are under constant pressure to find new and exciting roles for their latest offerings.

This has led to whole industries – photography, video, audio – being effectively annexed by the electronics industry. This is not necessarily to those industries' benefit: the camera or audio manufacturer's role can be reduced to 'badge-engineering', to assembling components and sub-assemblies made higher up the 'food chain', by the electronics firms. Their only alternative is to leave the industry altogether.

Obsolescence and indispensability are two sides of the same coin. Online shopping requires a reasonably up-to-date computer; but it also becomes increasingly necessary to shop online as that's where the best bargains are to be had (because 'everybody now shops online'), and certain kinds of items become harder to find anywhere else. So electronic devices become obsolete at the same time as their ownership becomes indispensable in a self-reinforcing feedback loop. The same goes for finding and applying for a job, keeping in touch with your friends, and even finding a phone number (printed phone directories having quietly faded away).

Buying a computer becomes less like acquiring an asset than incurring an ongoing liability that wasn't there before, but which one can less and less avoid as time goes on: a positional phenomenon.

Computers could be used to increase the size of the 'economic pie' for everyone while reducing human environmental impact. This was the confident expectation in the 1970s – but it would have required a large-scale, determined commitment to mutuality that failed to develop, and which is extremely hard to achieve in a society that has turned inter-personal rivalry into a cardinal virtue. Instead, the technology becomes a tool for appropriating bigger shares of the pie that already exists, by those who have the power to do so. Hence the lack of real economic growth identified by Robert Gordon, described in the last chapter. Industries and businesses are forced to look at each other as possible sources of food, and invest their effort accordingly. Computers have been used less to increase efficiency in everyday life than to concentrate economic power and wealth – which is disproportionately extracted from the poor and precarious.

THE RISE OF FINANCIAL SERVICES, TRAILED BY WOMEN IN OLD CARS

By augmenting businesses' ability to gain economies of scale, computers were a large factor in the consolidation into larger and larger units of retail, leisure and other public facilities that took place through the 1970s, 1980s and 1990s, and their migration to the outskirts of towns. The financial-services sector (a newly coined concept dating from this period) is a case in point.

In Britain, banking and finance were deregulated in the 'Big Bang' of October 1986 (whose consequences are explored further in Chapter 7). Rules separating investment and retail banking were abolished. Trading in the City of London was opened up to international business. Face-to-face trading was replaced by electronic, screen-based trading. Computerization made it possible to turn mortgages and pensions into tailored commodities (or at least, seem to do that) and sell them as lucrative alternatives to 'old fashioned' repayment mortgages and state and job-based pensions. Slick selling assumed paramount importance, and in many cases this meant investing in new, prestige headquarters, located out of town to give an enlarged sales-force easier access to motorways and airports.

A financial organization where I worked in the mid-1990s moved its headquarters out of the center of a lackluster town into a new, steel-and-glass atrium-style building, amid landscaped car-parks, close by the motorway. This meant that several hundred low-paid and mainly female clerical, catering and ancillary staff could no longer travel to work easily by bus, or do family shopping in their lunch-hours. So, over a period of three to four years, the car parks steadily became fuller and fuller – mainly with fairly old, low-status cars like Vauxhall Astras and Ford Fiestas and Escorts, which these poorly paid women had somehow managed to afford. I once overheard two executives complaining that their car-parks were being filled up with 'bangers'. Tennis courts and flower-beds were paved over to make way for them. Even so, it became necessary to arrive at work earlier and earlier to be sure of getting a place to park. Low-paid staff had been coerced into the automobile economy, to play their part in the ramping-up of carbon emissions that marked that decade.

This turned out to be part of a national and even global trend. Oxford geographer Amanda Root examined UK car-usage statistics for the 1980s and 1990s and found that the numbers of women with driving licenses rose by 90 per cent. For the first time there were as many female drivers as men (there had been only half as many in 1975-6); but the women only drove one fifth as many miles as men did[16] – reflecting the same car usage pattern that I had seen at first hand.

Work of all kinds has been relocated in a similar way. In her influential 1997 book *The Death of Distance*, Frances Cairncross assured her readers that the internet would bring people's work closer to where they lived, 'saving time and resources, and improving the quality of life'.[17] Yet, even as she was writing, work was moving further away, and had been for some time. More time was being spent in cars, creating a need for more reliable and safer models, leading, for example, to a 20-per-cent increase in the size of automobiles in the US between 1985 and 2007,[18] plus a tripling of commuting time in the 20 years between 1983 and 2003. The UN's *State of the World's Cities* report for 2008/9 found that US commuting times had increased again by a comparable amount (236 per cent) in the 10 years to 2009 (the actual miles driven in that time, however, had only risen by 25 per cent).[19] It is all of a piece with the intensified, long-distance traffic in goods and raw materials described in Chapter 4, adding to an already huge carbon footprint.

PUTTING A GIRL ON THE MOON: THE COST OF EDUCATION

Education does not look like an energy-intensive activity but social competition turns it into just that, and negates its purpose at the same time. In 1977, Fred Hirsch observed that education had already become a resource-hungry activity aimed more at providing individuals with saleable credentials, than at improving society's overall levels of wisdom and skill.

Hirsch was particularly interested in the way entry requirements for 'good jobs' (such as that of lawyer or doctor) had risen since the expansion of secondary education after the Second World War.

Whereas young people of elite parentage could more or less walk into Oxford or Harvard in the 1950s, and party their way through to some kind of a degree and then a plum job, their children found themselves up against competition from state-educated children, and had to take their education seriously. Parents started pouring more and more money into extra and better tuition and a sort of educational arms race commenced. Top grades all the way through one's school career became obligatory, and at least a 2.1 degree, as well as all sorts of extra-curricular accomplishments to make one's CV look more interesting.

Fred Hirsch called this the 'absorption of additional resources in credentials-producing educational activities' and observed that although it had gone through the roof, it had produced only a marginal increase in the numbers of doctors or their quality. (It may in fact have reduced their quality. More intense competition for the job of surgeon is thought to attract the exact types of people who should not be surgeons[20]). As we will see below, this also has a measurable and large effect on total human environmental impact – and thanks to our economies' continuing commitment to inequality, it has grown at an even faster rate since computer technology arrived on the scene.

In her 2012 book *Plutocrats*, the Reuters news-agency researcher Chrystia Freeland cited research showing that 'the wage premium for a college education [in the US] increased from

0.382 in 1970 to 0.584 in 2005, an increase of more than 50 per cent'; in other words, getting a college degree of any kind added an extra million dollars to a person's lifetime earnings.[21] But this conceals huge discrepancies: much of the premium goes to those who attend elite institutions. The Economist has reported that 'For grand places like Caltech and MIT, the 30-year return on a bachelor's degree is around $2m. But attending institutions near the bottom of the list actually diminishes earnings. Graduates of Valley Forge Christian College can expect to be made $148,000 worse off for their trouble.' Nonetheless, families feel they must invest in higher education in some form, for fear their children will end up even worse off.[22]

The cost of getting a child into an elite institution is extreme, although we hear about it through striking anecdotes that lack detail on the physical impacts. Freeland mentions an extremely wealthy Wall Street analyst who was so anxious about his daughter's future that he falsified a major company's share valuations so as to get his two-year-old child into the same nursery school as the child of the company's boss.[23] As Freeland explains, elite education at this level leads overwhelmingly to careers in finance rather than more productive areas of work, and is tightly integrated with the carbon-intensive lifestyle of the wealthiest 0.1 per cent, where private jets are the norm, along with multiple homes and offices in two or three different continents.

The physical impact of educational competition becomes more apparent when we look at how it affects ordinary wage earners' budgets. In 2002 Steve Gibbons and Stephen Machin worked out what educational competition among UK primary schools was costing them, in terms of house prices (a quintessentially unproductive expense that wage earners nonetheless have to pay). They found that 'a one-percentage-point increase in the neighborhood proportion of children reaching the target grade pushes up neighborhood property prices by 0.67 per cent'.[24] In 2006, the Royal Institute of Chartered Surveyors calculated that parents were 'willing to pay a £16,000 premium to live near a good school'.[25] By 2014 the premium had risen to £21,000 on average, and approaching half a million pounds in some areas.[26] These differentials flow from delegating the allocation of educational resources to inter-family competition. They represent a huge increase in what is shrewdly termed 'economic activity': activity that can be measured in cash terms, irrespective of what the activity actually achieves.

One can be certain, however, that this competition caused people to generate an impressive amount of carbon dioxide for marginal benefit, or just in the hope of not falling behind in the struggle for an acceptable standard of life: a purely positional phenomenon.

The struggle becomes self-sacrificial – and resource-intensive – among the least advantaged. I was told not long ago of a taxi-driver, from a largely South Asian inner-city area of Bristol, who had worked double shifts since his daughter was two years old to get her into and through one of the city's five private-sector all-girl secondary schools. He did this not only because he wanted his daughter to do well, but also because he feared she would sink without trace if she went to the local state secondary school, which was said to be 'rough'. He had to make this choice because recent UK governments have encouraged a private educational sector to flourish and, in wealthier places like Bristol, the trend has drawn funds and social commitment away from the state sector.

It might be supposed that major investment in elite secondary education would at least produce better educated people, but in 2011 Danny Dorling found that Bristol, with its huge proportion of private schools, was sending slightly fewer of its children to university than the similar-sized northern English city of Sheffield – which had a negligible private sector.[27]

This taxi-driver's choice was one that he and his daughter would be very much better off without. For example, a study published in 2011 suggests that he was over 50 per cent more likely to have a heart attack – simply as a result of working all those extra hours.[28]

But what about the environmental impact of his decision? A rough-and-ready calculation[29] suggests that, over the 16 years between his daughter's 2nd and 18th birthdays, his extra daily shift might have added more than 360 tonnes of CO_2[30] to the atmosphere. An economist might say that this should be set against all the extra, wealth-creating activity that taxi drivers make possible, in terms of getting people to appointments and so on. The burgeoning taxi population has itself been seen as an indicator of economic health (and it grew by 31.2 per cent in England, during the decade 2005-15[31]). But when we look at the kinds of journeys taxi users actually make nowadays, the picture changes completely. Taxi journeys are, increasingly, made by poorer people who do not have cars, and they use them for shopping, entertainment and hospital visits that could previously be made much more cheaply by public transport or on foot – before the iron laws of profit and loss extracted those facilities from town centers and suburban high streets, and concentrated them in out-of-town sites.

A 2011 survey carried out in Sydney, Australia, found that only 24 per cent of all taxi journeys were work-related; the overwhelming majority (52 per cent) were for 'recreation (such as entertainment, social visits, "going out", including getting back home)'.[32] A 2014 study by the Institute for Public Policy Research (IPPR) found that the heaviest users of taxis in the UK were now the poorest fifth of households – and they were using taxis mainly for want of adequate public transport.[33]

There is still the argument that taxi-driving at least offers a 'leg up' to unskilled people from the Global South, whose old ways of existence are no longer viable – but this, too, does not stand up against the actuality. A friend of mine turns out to be fairly typical: he is a medical scientist who came to Britain as a refugee from Iraqi Kurdistan. Apart from a few months as a temporary research assistant, the only steady work he has been able to find in his 10 years in Britain is driving taxis, and many of his colleagues have similar stories.

It is the same in other wealthy countries, and reflects a very general phenomenon: 'over-skilling'. A 2013 report says that as many as half of people who drive taxis in rich countries hold degrees, including doctorates, and a third of people admitted to the UK as 'highly skilled migrants' end up doing unskilled work.[34] In 2012, the UK's Office for National Statistics reported that '35.9 per cent of those who had graduated from university in the previous six years were employed in lower-skilled occupations...

THE BLEEDING EDGE

This compares with 26.7 per cent, or just over one in four, in 2001.'[35]

The process that insidiously changed education from a means of personal and social development into a 'credentials-producing activity', designed to help job applicants out-bid each other, is the same one that sets medical scientists to work driving taxis and fills their taxis with poor people who can't really afford the fares, but have to. This is what happens when there is societal tolerance of rising inequality, and political fear of confronting it, in a time of radical technological advance.

HOW 'E-LEARNING' REBOUNDED ON THE POOR

At the start of the electronics revolution there was enormous enthusiasm for computer-based learning. Seymour Papert's Logo computer language and his 1980 book *Mindstorms: Children, Computers and Powerful Ideas*, were emblematic of the benign and almost made-in-heaven relationship that was felt to exist between the new technology and education.

Education became an important area for the nascent industry (perhaps not least because, like defense, education is normally funded by governments). Britain's Open University (OU, set up by a Labour government in 1969) exemplified a possible development path. The OU used TV and radio, and then videotape and home computers – notably the BBC microcomputer, launched in 1981, which became an important ingredient in British computing culture in the 1980s and eventually led to the ARM microprocessors that now power so many mobile phones. The OU is now one of the world's biggest research and teaching universities, yet its purpose was emphatically not to be a 'credentials-producing activity'. Instead, its aim was simply to give higher education to whoever wanted it, whatever their age or ability, wherever they lived – not to provide skills considered necessary for the economy (although it turned out to be very successful at doing that).

The OU ethos had ceased to be dominant by the 1990s, when the terms 'learning economy' and 'lifelong learning' entered use – along with a more instrumental attitude to education. John

Schmitt, chief economist of the Washington-based Center for Economic and Policy Research, argued in 2009[36] that 'lifelong learning' played an important part in justifying increased inequality and the replacement of secure, well-paying jobs with lower-paid, less secure and usually de-unionized ones – for example, in call centers, which became a major global phenomenon from the late 1990s.[37] Ursula Huws and colleagues say that the occupational category 'call-center worker' first appeared in UK official statistics only in 2000 but calculated that there were already more than a million of them by 2003, and that call centers experienced phenomenal growth in virtually every country around the world in the first decade of this century.[38]

Organizations such as the OECD started to explain the rise of income inequality as the result of lower-paid workers' alleged failure to improve their skills to meet the requirements of the high-tech age. Schmitt wrote that this explanation was attractive because:

> [it] removed policy, politics, and power from the discussion of inequality, by attributing rising economic concentration to 'technological progress', a force that could be resisted only at our peril. The skills-biased technical change explanation also put significant limits on the terms of policy debates: the problems of the three-fourths of the US workforce without a university degree were either the result of the poor personal decision not to pursue enough education, or, at most, a sign that, as a society, we needed to invest more in education.

This explanation became conventional wisdom, promoted in magazine articles, and in the new genre of creativity and management books. Guy Claxton, author of *Wise Up: The Challenge of Lifelong Learning* put it bluntly: 'tomorrow's poor will be those who have not learned to keep up'.[39] In practice, this could mean keeping up with totally unnecessary changes created by obsolescence. For one IT worker interviewed by Chris Benner and his colleagues in their 1999 report, *Walking the Lifelong Tightrope*, 'lifelong learning' meant a psychologically and financially ruinous process of 'having to constantly "learn" the same old thing over and over'.[40]

Higher education became a particular target for IT firms. With the concept of 'distance learning' came the idea that universities could 'expand their markets' without expanding the numbers of teachers (or even while reducing their numbers, or getting more 'value' out of each teacher – as described by historian David Noble in his 2001 book *Digital Diploma Mills: the Automation of Higher Education*[41]). Government ministers (who usually knew very little about computers) made suggestible subjects for corporate lobbying.

The new world of computers offered a seductive option to politicians daunted by the growing moral challenge of inequality: blame it on the failure of 'those less equal' to keep up with the technology, and let the electronics industry ride to their rescue with computerized solutions. From the employee's or would-be employee's point of view, computers, and learning how to use them, became an essential cost of the business of 'keeping up'.

The billions spent on e-learning systems in schools and in universities did not radically improve the availability of many important skills. Instead of training and employing more doctors, teachers, care workers and so on, the less contentious route was chosen, of targeting and weeding out inefficiency and extracting more useful work from existing employees. Computers lent themselves all too easily to this task, eroding the security and autonomy that had made many jobs desirable and which often, in the eyes of the workers themselves, were essential to doing the jobs well.

At the confluence of these trends we see a 'bow wave' of wasted and unpaid human effort: the rising amount of work required simply to get a qualification, and then a job (or benefits, if you are unable to find work or cannot work).[42] Thanks to the power imbalances that guided the project, the thrust of development as the technology advanced became more and more about testing and checking on people, and less and less about empowering them – defining the future character of the technology in the process.

There are no good reasons why there should ever be a shortage of 'good jobs', other than lack of political will. The more good doctors, teachers, plumbers and so on, the better. Numbers of doctors per 1,000 patients vary considerably from country to country, with the more egalitarian (and lower-impact) countries tending to provide more than the less equal ones. Whereas the US and UK had only 2.4 and 2.7 doctors per 1,000 patients in 2010, France had 3.4, Denmark 3.5, Sweden 3.8, and Norway 4.2. Cuba, with one of the world's most egalitarian societies, provided 6.7 doctors per thousand of its population.[43]

Doctors in Cuba are paid a lot less than they are elsewhere, yet the profession still has no difficulty attracting recruits – and in 2006 Cuba achieved slightly better health outcomes for its people than the US did, for just under a 20th of the expenditure per patient.[44] Cuba has regularly been able to lend doctors abroad to help with crises and to build up local services, providing 'more medical personnel to the developing world than all the G8 countries combined'.[45]

Cuba's success has been helped by a different approach to e-learning. During the 'special period' in the early 1990s, when the Cuban economy faced collapse after the fall of the Soviet Union, an international award-winning, Linux-based e-learning and knowledge-sharing system (INFOMED) was created to empower health workers and disseminate information. The system treats qualifications as a side-issue. The main focus is the sharing of information between clinics, and between workers of all grades, in order to identify emerging health trends and diffuse good ideas rapidly.[46] The contrast in approach with the e-learning trends outlined above could not be more stark.

1   George Marshall's book *Don't even think about it!* Bloomsbury, 2014, exemplifies this tendency.
2   William Stanley Jevons, *Political Economy*, 1880, Chapter 2.
3   Thomas Piketty, *Capital in the Twenty-First Century*, Harvard University Press, 2014, pp 509-510, and notes on p 638. During the presidential election campaign of 1972, Richard Nixon's opponent, the Democrat George McGovern, was proposing a top tax rate of 100 per cent on inherited wealth.
4   Andrew Revkin, *The Burning Season,* Houghton Mifflin, 1990.

5   O2 'Young Brits Take Mobile Attachment to the Extreme', nin.tl/O2young Accessed 27 Mar 2014.
6   T Kelly, 'Mobile phone credit instead of bread? For many Kenyans, a real dilemma', 2012, nin.tl/mobileorbread Accessed 28 March 2014.
7   RH Frank, *Falling Behind,* University of California Press, 2007.
8   According to Wikipedia, the phrase comes from a character in the *Babylon 5* TV series, broadcast in the early 1990s.
9   Stephen Bayley, 'Let's start thinking outside the box', *The Observer*, 15 July 2007. He refers to preferences given by home owners in a study by the Royal Institute of British Architects: 'No more shoddy Noddy boxes'.
10 Piketty, op cit, p 503.
11 M Glaskin, 'Shockwave traffic jam recreated for first time', *New Scientist*, 2008, describing the work of Y Sugiyama, Minoru Fukui, et al (2008). 'Traffic jams without bottlenecks', *New Journal of Physics* 10, 2008. The video of their demonstration, which uses cars on a circular test-track, is available on the web.
12 D Helbing, 'Traffic and Related Self-Driven Many-Particle Systems,' *Reviews of Modern Physics* 73, Oct 2001.
13 J Turley, 'The Two Percent Solution', 2002, nin.tl/TwoPercent
14 Anonymous, 'Tantalum Capacitors: Global Trends in 2000', *Passive Component Industry*, Jan/Feb 2000, pp 24-27.
15 J Turley, 'Embedded Processors by the Numbers', 1999, nin.tl/TurleyEmbedded.
16 Amanda Root, 'Transport and Communication', in *Twentieth Century Social Trends*, ed Arthur Halsey and Jo Webb, St Martin's Press, 2000.
17 Frances Cairncross, *The Death of Distance*, Harvard Business School Press, 1997, p 217.
18 Richard Frank, 'Falling Behind: how rising inequality harms the middle class', University of California Press, 2007. The weight of a Honda Accord (an average car) increased from 2,500 pounds in 1985 to 3,200 pounds in 2007.
19 State of the World's Cities 2008/9: Harmonious Cities, Taylor & Francis, 2012.
20 For example, see Irina Shlionskaya, Plastic Surgery, nin.tl/Shlionskaya
21 Chrystia Freeland, *Plutocrats,* Penguin, 2012, p 48.
22 'Wealth by degrees', *The Economist*, 28 Jun 2014, nin.tl/wealthbydegrees
23 Freeland, op cit.
24 Steve Gibbons & Stephen Machin, 'Valuing English Primary Schools', July 2002, nin.tl/valuingschools
25 'Parents pay on housing for a "good school"' *The Guardian*, nin.tl/parentspay
26 Jonathan Owen, 'Parents Pay Half a Million in Housing Premium for State School Education,' *The Independent*, nin.tl/housingpremium Accessed 26 Aug 2014.
27 Danny Dorling, *So you think you know about Britain?* Constable, 2011.
28 Medical Research Council News, 'Working 11-hour days increases your risk of heart disease by over 50%', nin.tl/11hourdays
29 Assuming a normal shift involves driving 1,500 km per week (75,000 km per year, if he works 50 weeks out of 52), and this man's taxi averaged a

typical 8 kilometers per liter of diesel, he would get through 9,377 liters per year. According to the Belgian website Ecoscore, each liter of diesel produces 2.64 kilos of CO_2, – making a grand total 24.75 tonnes of CO_2 in a year for a taxi working normal shifts.
30 Carbon figures from nin.tl/carbonfigures ; Fuel consumption figures from nin.tl/1Lyko6d ; The 1,500 km/week figure was supplied by a taxi-driver.
31 UK Government Statistics, 'Taxi and Private Hire Vehicle Statistics England 2015', nin.tl/taxistatistics
32 Taverner Survey Report: 'Taxi Use Sydney', Nov 2012, nin.tl/taxiuseSydney
33 Mark Rowney & Will Straw, 'Greasing the wheels', Institute for Public Policy Research, 26 Aug 2014, nin.tl/IPPRbuses
34 S Pegiou, 'Over-qualification of immigrants', *Migration for Development*, 2013, nin.tl/migrantbraindrain
35 Graham Snowdon, 'A third of recent graduates in unskilled jobs', *The Guardian*, 6 March 2012.
36 J Schmitt, Inequality as Policy? Oct 2009, nin.tl/inequalitypolicy
37 David Holman, Rosemary Batt & Ursula Holtgrewe, Global Call Center Report, 2007, nin.tl/callcenterreport
38 Ursula Huws, Simone Dahlmann, Jörg Flecker, Ursula Holtgrewe, Annika Schönauer, Monique Ramioul & Karen Geurts, Value Chain Restructuring in Europe in a Global Economy, Katholieke Universiteit Leuven, HIVA, 2009.
39 Guy Claxton, *Wise Up: The Challenge of Lifelong Learning*, Bloomsbury/St Martin's Press, 1999.
40 Chris Benner, Bob Brownstein, Amy B Dean, *Walking the Lifelong Tightrope*, Working Partnerships USA, Economic Policy Institute, 1999.
41 David F Noble, *Digital Diploma Mills*, Monthly Review Press, 2001.
42 As a small indication of the rate at which this kind of work has been increasing, the Randstad employment agency reported in 2013 that in the previous four years the amount of time people were spending on getting a new job had risen from just over 8 weeks to over 10 weeks. See: Louisa Peacock, 'Jobhunters spend over 10 weeks searching for new job', *Telegraph*, 26 Feb 2013, nin.tl/10weeksearch
43 Data drawn from WHO by World Bank at data.worlldbank.org Accessed 3 June 2015.
44 On numbers of doctors and health expenditure, see Guardian Fact File UK, 6: Health and Food, 29 April 2010; On life expectancy, see WWF Living Planet Report 2009 et passim; on Cuban doctors, see Wikipedia entry on Cuban Medical Internationalism (accessed 5 Aug 2011).
45 Robert Huish and John M Kirk, 'Cuban Medical Internationalism and the Development of the Latin American School of Medicine', *Latin American Perspectives*, 34; 77.
46 Ann C Seror, A Case Analysis of INFOMED, Cité Universitaire, Quebec, 2006, jmir.org/2006/1/e1

7
Enclosure in the computer age: the magic of control

Intellectual property law is the modern counterpart of the brutal land enclosures of previous centuries. Computers and the internet have brought managerial control, commodification and rent-extraction into the most intimate areas of life and work. Finance shifted from productive investment to investing in debt – and collapsed in 2007/8. Meanwhile ordinary people's worlds shrank. But other worlds are possible. At the heart of the storm, computer programming remains an old-fashioned craft industry, defying all efforts to eliminate it.

As inequality increases, those who enjoy the unequal share of power use it to extend their control of the possibilities for wealth creation. Rent (income one enjoys as reward simply for owning something) becomes more lucrative than any wealth one could possibly achieve by making something. Even simply possessing an option to make something can be more lucrative, and certainly less troublesome, than exercising that option. We have seen this tendency let rip in the decades since capitalist firms overcame their original reluctance to use computers.

In *Capital in the Twenty-First Century*, Thomas Piketty pointed out that, in an unequal society, entrepreneurs tend overwhelmingly to turn into rentiers. Even supposed exemplars of creative capitalism, such as Bill Gates, derive vastly more of their income from rent (fees derived from intellectual property) than they ever did from the creative work on which their businesses are based.[1] Creators who fail to assert intellectual property rights need to be content with vastly fewer material rewards (as is Daniel

Bricklin, who invented the spreadsheet two years before US law was changed to allow general ideas such as his to be patented).

The latest expansion of rent-extracting powers by the 'haves' has been described as 'the new enclosures'.[2] The comparison is often made in a slightly tongue-in-cheek way, as if the brutal land enclosures that took place in England in the 16th century and peaked in the early 19th bore only a metaphorical resemblance to the present expansion of intellectual property law. But the dynamics are the same and the effects every bit as destructive – and they operate on a global scale. Old and new enclosures both followed from sustained lobbying by elites who at first enjoyed only slight success, but then, as their power grew, became able to dictate the political agenda and enjoy state aid with its enforcement.

We easily forget that the capitalist concept of the basic means of production, the land (as someone's property, with a fence round it) is a recent and very odd part of the human story. As late as 1902, Petr Kropotkin was able to describe region after region in Europe and Russia where collective ownership and management of land continued to be the rule, the particular self-management methods in each case, and their often striking achievements (for example, a radically more efficient plough, developed in the 1890s by a network of peasant communities in southern Russia).[3]

The sheer shock of encountering the capitalist approach to the land was described by witnesses to the formation of the early North American colonies. The people of the Mohawk basin (now upstate New York) felt there was plenty of land for everyone and were generous with it – but were startled and then outraged when the Europeans put fences around their plots to keep others out, even when they were doing nothing with it.[4] George Washington built up his fortune this way, buying and importing bonded paupers from England to fence and defend the 60,000 acres he acquired in the Shenandoah and Ohio valleys, even before the territories had been formally colonized. His biographer Joseph Ellis ascribes Washington's extraordinary 'appetite for acreage' to a pathological obsession with status and terror of losing it. Later on, from 1784, he fought and won a two-year court case to evict a group of poor settlers who had innocently supposed the land to be unclaimed.[5] Similar scenarios have been enacted all over the colonized world.

Enclosure is good for profits but bad for productivity. Over centuries it becomes normalized but the shocking fact was still fresh and self-evident in England in 1516, when Sir Thomas More saw land that had supported thriving human communities turned over to sheep. In *Utopia*, he wrote:

> Your shepe that were wont to be so meke and tame, and so smal eaters, now, as I heare saye, be become so great devowerers and so wylde, that they eate up and swallow down the very men them selfes. They consume, destroye, and devoure whole fields, howses and cities... Noble man and gentleman, yea and certeyn Abbottes leave no ground for tillage, thei inclose all into pastures; they throw down houses; they pluck down townes, and leave nothing standynge but only the churche to be made a shepehowse. [6]

Positional considerations can militate in favor of keeping land out of production altogether, and it is not just wealthy landowners who do this.

Robin Jenkins (who farmed in Portugal in the 1970s, as the country began to modernize after the Salazar dictatorship) noticed how peasants who had found factory work around Lisbon airport hung onto their little pieces of land as insurance, in case their relatively well-paid jobs disappeared. The land, which had been wonderfully productive before, generally went to weeds.[7] At the other end of the wealth spectrum, 45 per cent of Cambodia, including nearly all its coast, disappeared in 2007/8 into the hands of overseas investors, who suddenly needed alternative ways of investing their money after the global financial crash. As *The Guardian* reported, most of this land was then fenced off and did nothing; a few people continued to work there, as security guards.[8]

Strengthened intellectual property laws are argued for nowadays on the grounds that they supposedly lead to higher productivity. The same argument was used to justify enclosure of land – but this was easily exposed as a fantasy, even in the place it was assumed to have been most successful: 18th/19th-century England. Economic historian Robert Allen has shown that the big gains in agricultural output during the Industrial Revolution owed little to enclosure. Peasant farmers' yields increased at nearly or the same rate, as they adopted the newer methods (something peasants were supposed not to do).[9] Any inferiority in yields was more likely due to relative lack of capital. And long ago, in the 1920s, the pioneering social historians John and Barbara Hammond had showed that the promise of increased yields was often mere cynical propaganda-cover for driving recalcitrant but productive peasants off the land. Since Otmoor, near Oxford, was violently cleared of its people in 1815, it has produced almost nothing[10]; it is now a nature reserve, yet its biodiversity remains lower than when it also supported hundreds of humans.

As Kropotkin said, 'to speak of the natural death of the village communities in virtue of economical laws is as grim a joke as to speak of the natural death of soldiers slaughtered on a battlefield'.[11]

These transformations were effected by people a long way away who often did not bother to go and look at what was happening, or even inspect their new property. That kind and degree of power, to transform a landscape from a distance, and make people vanish, had only happened previously in fairy stories.

THE SUPERNATURAL ENTERS EVERYDAY LIFE: THE MAGIC OF COMMODITIES

Positionality sheds light on the logic, and the magic, of commodities: phenomena that are now so omnipresent that we fail to notice their sheer oddness.

From a capitalist's point of view, a commodity is a pure, idealized, clearly defined, standardized, desirable something, with the potential to be summoned up in any quantity, the same every time. Hence, a printed book, or a plastic toilet seat, or an antibiotic pill, or a bullet is a commodity. Natural products like potatoes, timber, iron ore and grain can also be treated as commodities, after a fair amount of work has been done to standardize them – so that, for example, the buyer for a large catering firm can pick up a phone and order 'potatoes' or 'carrots' in any quantity he needs, at a suitable price, without any fuss about their particular provenance or qualities, and switch from one supplier to another quickly.

As Karl Marx says in the section of *Capital* that discusses 'the fetishism of commodities', a commodity is 'a very queer thing, abounding in metaphysical subtleties and theological niceties.'[12] And the queerest, most metaphysical thing of all is its price. The price is acutely, agonizingly sensitive to opinion but, whatever the commodity's other qualities, in the last analysis, that is the only thing about it that matters.

Somewhere, the commodity's price has been described as its 'little, beating heart', like that of the fairy Tinker Bell in JM Barrie's *Peter Pan*, which will stop beating if children stop believing in fairies. Maintaining a product's credibility really is a matter of life and death, and it absorbs a great deal of a capitalist's energy and resources.

The ideal commodity is one that can be turned out in any quantity as if by turning a handle, and which the populace will line up eagerly to buy as fast as it can be sold. As we will see in Chapter 8, microchips have proved to be something dangerously close to being the ideal capitalist commodity.

There is naturally intense competition to control the means to produce commodities. When it becomes possible to create new means of producing commodities (for example, machines) there is also intense competition to own the ideas that define them, as well as to decide what gets produced, who turns the handle, how fast, and so on. In this fashion, 'intellectual property' turns thought itself into a positional good. If I have an idea and patent it, you cannot have it, even if you think you can.

Work becomes a positional good (a scarce source of paid employment, for which people compete) and a commodity (precisely defined, and in an ideal capitalist world the people who do it are also interchangeable commodities). The whole set-up is a power landscape, with pinnacles and mountains from which the lucky few direct operations.

POWER OVER THE FUTURE: THE MAGIC OF INTELLECTUAL PROPERTY

Commodification goes hand in hand with intellectual property (IP: patents, trademarks, copyright), enforced by national laws and international treaties. The value and power of IP have soared since the computer's entry into the capitalist economy, especially since the 1980s, in terms of the range of things to which IP can be asserted, as well as its importance to Western economies. This is in part because more and more of the procedures that once lived in workers' heads and hands could be replaced by ones that were written down, and therefore capable of becoming someone's property.

This was largely an effect of the tendency described in the last chapter, for the power of business lobbies to grow as wealth shifts in their direction, so that they can persuade law makers to shape the law even more in their favor. A famous example: the Walt Disney company lobbied successfully in 1998 to have the US copyright term extended to 100 years, to prevent the copyright on their lucrative Mickey Mouse character expiring in 2003. This had far-reaching effects on the US entertainment media industry: its subsequent effort concentrated much more on the development of franchisable characters and formats, and a preference for sequels rather than new stories.[13]

In June 2006, it was reported that the 'intangible' assets of US firms were worth around $6 trillion – two-thirds of their total value. Much of this consisted of IP in the form of patents (whose licensing raised $100 billion a year) and trademarks that had become so valuable that they supported a rapidly growing bond-market of their own, valued at around $106 billion. When the Dunkin Donuts brand changed hands that year, its new owners were able immediately to recoup the $1.7 billion cost of their purchase by issuing bonds backed by the brand. 'Securitizing' intellectual property in this way raises its importance still further, as a key source of investment capital.[14]

Intellectual property has become one of the world's strongest-growing commodities: according to the World Trade Organization, licensing fees and royalties made up 6.4 per cent of world exports in 2011 and were worth $270 billion, having grown by 13 per cent since the previous year. Nearly all of this went to the US and Europe ($108 billion and $115 billion respectively).[15]

However, its effect on what people pay for things (and therefore on the economy that depends on their paying as much as possible for things) is very nearly 20 times greater. According to the US Patent and Trademark Office, in 2010 'IP-intensive industries' were worth $5.06 trillion to the US economy, and a similar amount to that of the European Union. They employed 27.7 percent of all US workers – and these were increasingly better paid.[16] In terms of patents, the top four of these industries were all in the 'computer and electronic product manufacturing' category, with pharmaceuticals and chemical products coming in just behind them. In terms of copyright and trademarks, the motion picture and video industry dominated. The category 'Lessors of nonfinancial intangible assets' – firms that administer licenses, overseeing branding- and rights-deals and so on – earned $24.5 billion in 2007, but employed only 25,500 people in 2010: a good business to be in.

The US-based sports-shoe company Nike illustrates another aspect of the rich countries' retreat from actual production to rent-extraction via IP. By the late 1990s, Nike had not only moved production but also its design, accounting, shipping and marketing to its network of major suppliers and investors in the various states and special regions of the 'Greater Southern China Economic Region' (GSCER). Nike's own role was redefined as 'brand proprietorship'[17] so that the company can take an even greater share of the proceeds from its products, while contributing less and less to their creation.

Thanks to computer-aided design (CAD), which produces data that can be owned, IP can be asserted all the way down to the point where needle touches cloth or leather, irrespective of who owns the sewing machine, or did the design. The same system makes it possible for fashion clothing manufacturers to specify, get quotes for, and outsource separately every bit of fabric, zip-fastener, button, stitch and seam of a garment, even as it is being designed, and fine-tune the design to balance its appearance with its profitability, squeezing the supplier's already-slim margins to the limit.

The computer-aided design and manufacturing (CAD/CAM) systems that are used to design garments even have access to libraries of 'Standard Minute Values' for each operation, such as putting a dart into the waist of a blouse, or adding a patch-pocket, which makes it easy to calculate the precise amount of labor-time the whole garment 'should' require. Almost nothing is left to the supplier's discretion, and less is required from the suppliers in the way of expert advice – all that's wanted is a price, and a low one. The lists of cutting and sewing instructions are emailed to suppliers in Vietnam, China, Cambodia, Thailand, Indonesia or Malaysia (or some poor enclave in a wealthy country) who then 'blind tender' for the work, hoping that their quote will be cheapest. Importers can, and do, switch manufacture across continents at a moment's notice.[18] The result is massive under-pricing by anxious suppliers, with the costs borne by mainly female workers.

In ways like these the market has extended itself, almost as a by-product of the computer industry. After all, it was electronics firms that pioneered from the 1980s onwards the practice of shifting jobs and then entire plants and processes, swiftly, to wherever work can be done most cheaply, and then designing jobs and processes so that they could be relocated in this way. By 2000, electronics was 'the most globalized of all industries'[19] and computers and the internet were being used to globalize everything else as well.

IP DOESN'T WORK WITHOUT A STRONG STATE

Just how much this regime owes to the power and willingness of governments to enforce IP law is clear from the proliferation of inter-governmental agreements (starting with TRIPS – the Agreement on Trade-Related Aspects of Intellectual Property Rights – concluded by the World Trade Organization in 1994). Countries have to sign up to these agreements, and enforce them under their own laws, as a condition for credit, aid and trade deals.

The IP system's dependence on state power is also evident from the almost comical way the prices of patented goods fall the moment they leave the protection of Western IP law. To take one of innumerable possible examples, the leukemia drug Gleevec costs $70,000 per year in the US, but only $2,500 in India.[20] This is because India declined to sign up to the relevant international agreement. Under India's own IP laws, new processes are patentable, but not the products themselves – so if manufacturers can find other ways to make them, they can do so. This has made India literally a lifeline for millions of people who would never be able to afford Western prices for essential drugs.

Intellectual property rights are one more way in which unequal power ends up stifling innovation and the spread of ideas. Evidence of this is the rising importance of 'preventive' or 'submarine' patents – patents taken out by manufacturers on potential processes and products that, if they were to come to market, might damage sales of their own product lines. Enormous creativity and effort is poured into research, simply for the purpose of preventing anyone using its results. In their 2008 book *Against Intellectual Monopoly*, Michele Boldrin and David K Levine say that anything between 40 per cent and 90 per cent of all patents issued in the US were neither used by the patentees nor licensed to others; they were 'submarine patents', existing solely to protect an existing product-line from possible competition.[21]

If technology is a true evolutionary phenomenon, as economist Brian Arthur suggested (in Chapter 1), this is as if at some point in the early Cretaceous period, 145 million years ago, jealous gods with major interests in dinosaurs and tree ferns had patented mammals and flowering plants to prevent their further development – abruptly truncating the story of life on Earth 90 million years later, when an unexpected meteorite made the dinosaurs and tree ferns extinct. This isn't as far-fetched as it sounds: US companies have had the right since 1952 (under Section 102 of the US Patent Act) to patent traditional uses of life forms found in other countries (and therefore, in effect, the life forms themselves).[22] If the current legal regime had been in place between the 1940s and the 1990s, it would probably not have been possible to develop computers. As Donald Knuth, a revered authority on computer programming, wrote in an open letter to the US Patent Commissioner in 1994:

> When I think of the computer programs I require daily to get my own work done, I cannot help but realize that none of them would exist today if software patents had been prevalent in the 1960s and 1970s.[23]

This is what Bill Gates, founder of Microsoft, said on the same subject in 1991:

> If people had understood how patents would be granted when most of today's ideas were invented, and had taken out patents, the industry would be at a complete standstill today.[24]

Confronting the conundrum of his own dependence on progress-inhibiting intellectual property rights, he concluded that: 'The solution to this is patent exchanges with large companies and patenting as much as we can.'

Thinking back to 1946 (and Chapter 1), one imagines that John von Neumann would have had to sign a non-disclosure agreement with Eckert and Mauchly before looking at the ENIAC computer, preventing him from disseminating the computer specification on which the 'new economy' is built. We should be wondering what other technologies we might have had, if the Western countries' post-War interlude of egalitarianism and openness had been defended with a will.

IP law, like the original enclosures, is imposed by elite power, purely extractive in function and, as we begin to see, environmentally destructive.

COMPUTERS AND THE MAKING OF MONEY

Money would never be a problem if it were created and maintained by the community that uses it (which is the case in traditional societies, as David Graeber has explained,[25] and could be the case everywhere, according to Ann Pettifor[26]). But that would require equality or something approaching it.

Rising elite power achieved the 'liberalization' regime that enforced the intellectual-property regime described above. Its larger achievement was to grant commercial banks more and more power to define and create the money that people must use to buy commodities. This is done by issuing bank credit – 'stroke of the pen money'. Contrary to popular belief, and even some economists' belief, banks really do create money out of thin air when they give someone a loan or increase their credit-card limit. 'The process,' wrote Canadian economist John Kenneth Galbraith in 1975, '...is so simple that the mind is repelled. Where something so important is involved, a deeper mystery seems only decent.'[27]

Banks gained greater freedom to do this as power and wealth brought them greater lobbying power and the very latest computing science. It suited elites to have money created this way, rather than by raising wages. Moreover, consumer credit earns extremely good interest and, even more seductively, the computers appeared to make it possible to protect investors against defaults by securitizing debt according to sophisticated mathematics, and then to make even more money by trading the securities on the financial markets. In the US, bank credit rose from 21 per cent of GDP in 1980 to 116 per cent in 2007: this was the real basis of the economic boom over that period.[28] In the UK, bank credit rose from 36.2 per cent of GDP in 1980 to 136 per cent in 2000, and 212 per cent in 2008.[29]

By the 2000s, 80 per cent of the money in circulation consisted of bank credit, and easy credit had become a long-term substitute for wage increases. Household debt in the US rose to 70 per cent of disposable income in 1985, and to 122 per cent in 2006[30] – reflecting the growing wealth gap between rich and poor. Between 1982 and 2004, world financial assets increased 32-fold, while average incomes only rose 3-fold – barely keeping pace with inflation.[31]

The disaster that unfolded in 2007/8 was the result of harnessing the enormous power of computers to an entirely unfounded belief system, in which prices are the only signals markets need in order to regulate themselves. More and more powerful computer systems provided more and yet more information about prices, delaying but also amplifying the crash that eventually happened. Prices give no better idea of the state of a market than the speeds of nearby cars on a motorway do of the conditions ahead. Drivers can see each other's speeds just as easily as dealers can see the prices of commodities, but accurately predicting the likely changes in those speeds and prices is beyond the capabilities of any conceivable computing system.

This awkward fact had already been proved in 1963 by the French mathematician Benoit Mandelbröt, who is now best known for his discovery of 'fractals' (the snowflake-like, tree-like and other shapes found in nature, which have 'fractional dimensionality': neither one-dimensional line nor two-dimensional plane).

While working as a researcher at IBM, Mandelbröt analyzed the movements of the prices of bonds in the US cotton market over the previous hundred years. He found that prices could change dramatically and suddenly, in obedience to an entirely new class of equations, which helped to give rise to a new science known as Complexity Theory. As the German economist Tobias Preis points out,[32] Mandelbröt's work exposed the fundamental flaw in liberal economics – but this did not discourage others from trying computerized, mathematical approaches, which opened the way to an enormous expansion of derivatives trading, following the publication in 1973 of Fischer Black and Myron Scholes's algorithm for the pricing of options.

Before Black and Scholes's work, derivatives had been a small fraction of all investments, essentially the contracts known as 'futures', which offer the right to buy commodities like coffee, wheat and copper at some point in the future at a pre-agreed price. Futures are benign insofar as they help such industries keep their cash-flow steady, pay their workers and plan for the future. Black and Scholes's algorithm allowed almost everything else to be priced in the same way, including personal debt. It was what made the 'hedge fund' possible, and a world economy in which financial assets completely eclipsed the real ones that they supposedly represented. By 2007, the global derivative market was estimated to be worth $750 trillion compared to global output of $60 trillion.[33] The credit-default market in 2008 was worth another $55 trillion – roughly the same as the cash value of the world's entire productive output.[34]

The power-shift towards finance became self-reinforcing. Financial institutions deployed their increased lobbying power successfully against the legal barriers between speculative and retail banking (introduced in the US in the 1933 Glass-Steagall Act to protect against exactly this kind of speculation bubble, and repealed by the Clinton government in 1999; in Britain the barrier was demolished earlier, in the so-called 'Big Bang' of October 1986). This drew creative effort and investment into the monetary system itself, and away from the productive economy for which monetary systems in theory exist.

Scholes, and his and Fischer Black's collaborator, Robert Merton, were awarded the 1997 Nobel Prize in Economics for their work (Black had died the year before, and Nobel prizes are not awarded posthumously). The very next year, 1998, the company they had formed to exploit the algorithm failed and had to be bailed out with $4.8 billion of public funds. Further collapses followed, including that of Enron in 2001 until, in 2007/8, the whole market failed, after the collapse of the sub-prime mortgage market, which had been built on the Black-Scholes approach.[35]

Fischer Black's nephew, Michael Black, now a moral theologian at the Dominican Blackfriars Hall in Oxford, also rose high in corporate finance and witnessed the way his uncle's work played out there in the 1990s and early 2000s. He describes Black-Scholes as 'probably the closest the world has ever come to a universal standard of financial value', but only from the privileged vantage point of those who had the power to apply it. He says: 'the real money in corporate finance didn't lie in creating value but in defining it'[36] – and that is what Black-Scholes, plus computers, allowed an opportunistic financial elite to do for more than three decades.

The ability to define value, and to move money instantly to seize momentary opportunities and evade risks, led to a computer arms-race in the world's financial centers which brought their computers close to literal meltdown long before the figurative, financial meltdown happened in 2008 – as described by Donald MacKenzie in a study of the computer systems at Canary Wharf, hub of the British financial industry, earlier that year:

> The credit market is... one of the most computationally intensive activities in the modern world. An investment bank with a big presence in the market will have thousands of positions in credit default swaps, CDOs, indices and similar products. The calculations needed to understand and hedge the exposure of this portfolio to market movements are run, often overnight, on grids of several hundred interconnected computers. The banks' modellers would love to add as many extra computers as possible to the grids, but often they can't do so because of the limits imposed by the capacity of air-conditioning systems to remove heat from computer rooms. In the City, the strain put on electricity-supply networks can also be a problem. Those who sell computer hardware to investment banks are now sharply aware that 'performance per watt' is part of what they have to deliver.[37]

These events and discoveries have given some impetus to the new discipline that attempts to build an economics based on the one really 'hard currency' we know of: energy. We will touch on this again when we look at the energy cost of the kind of computers we have now, in Chapter 9.

THE WORLD GETS SMALLER AND HOTTER

All of this escalation and intensification of positional power has made the world much, much smaller – but not in the attractive way evoked by the phrase 'global village'. The aspects of computerization that have made the world more village-like have also provided cover for serious physical restrictions on people's movements which, perversely, intensify the energy costs of simply getting from A to B.

'In a more unequal society, everyone is less free to choose where they live', says Danny Dorling.[38] His 2007 study *Poverty, Wealth and*

THE BLEEDING EDGE

Place in Britain, 1968 to 2005 showed that even the winners in an unequal society are quite severely restricted as to where they can and can't go.[39] The 'exclusive rich' are to a large extent byproducts of computerization (overwhelmingly, their fortunes come from the finance sector, and from the high-tech industries that support it). Because they are so rich, and others are so poor, they must use more and more of their personal resources to compete for diminishing numbers of desirable locations, and to avoid the growing numbers of undesirable ones. Islands of respectability and safety become smaller and more isolated. Getting to them can be as serious a business, and as energy-intensive, as travelling around an occupied country.

At the same time, poorer people are confined to ever smaller localities which, as discussed in the last chapter, are also further and further from their work, so commuting times and emissions escalate. Living space is more and more constrained as ownership is intensified. In Britain, new houses have actually shrunk in size, at the same time as their prices have risen.[40] How small one's world can become was illustrated by a 2008 report from the UK's Joseph Rowntree Foundation, whose authors interviewed a representative group of young men who were effectively confined to a bleak area just 200 meters square on the fringes of the impoverished town of Peterborough.[41] They felt unsafe outside that area, and could seldom afford to go far beyond it.

The pressure works its way right to the bottom of the pile and more and more of the people there end up in prisons, which are a form of housing, after all (as Dorling points out in his 2014 book, *All That Is Solid*[42]). Not only are there more prisons in more unequal countries, with more of the people in them for longer; they are also more closely packed (two and three prisoners to a cell intended originally for one). Prisons are also some of the most energy-intensive buildings, according to a UK report in 2010,[43] and the transportation and other systems that support them amplify their impact.[44] A new sub-industry ('escorting services') provides the fleets of energy-hungry armored vehicles, known as 'sweatboxes', that are now familiar sights on British motorways, moving prisoners to and fro, the length and breadth of the country between courts, police stations, hospitals and prisons at all hours of the day and night. They are operated (as many prisons now are) by transnational outsourcing firms such as GEO, Serco, Sodexho, Wackenhut and G4S. Many of these new transnational giants rose from humble beginnings, cleaning washrooms and delivering school meals, with the trend to use computers to commodify areas of service work that had not been commodifiable before.

As a society becomes more unequal, risk increases. People must divert energy from productive and enjoyable activities to protecting themselves against possible future harm – which includes failure to seize opportunities as they arise, and inability to respond to threats. People become more vigilant at every level of society and levels of trust fall.[45] More and more resources are invested by individuals, firms and nations in their readiness to compete. As unequal societies become very unequal, powerful firms can externalize some of their 'readiness costs' by employing casual staff, or even putting staff on zero-hours contracts, so they do not even have the option of working for anyone else, and have to exist somehow at their own expense until and unless the employer needs them.

As inequality increases, societies invest more in the ultimate form of 'readiness to compete': up-to-date armies with lavish military training facilities and intensive weapons research. Computerization plays the same role here that it does in the financial sector, creating an illusion of infallibility based on ever more exhaustive data gathering, to serve ever fewer yet more powerful decision-makers: a cycle of enhancement that absolutely fails to predict real-life threats, yet which no general or politician dares to abandon.

Military effort is overwhelmingly carried out in the wealthiest states, which have most to defend; the Stockholm International Peace Research Institute (SIPRI) says that 'high-income countries account for about 75 per cent of world military spending but only 16 per cent of world population'.[46] Among these high spenders, the ones that have the highest expenditures relative to GDP are all highly unequal ones: the US (which alone accounts for 43 per cent of world arms spending), Russia and Saudi Arabia; the very-unequal

UK spends a slightly lower proportion of its GDP on weapons, but in 2014 it still ranked sixth in the world.[47] and was the world's second-largest exporter of armaments (just ahead of Russia).[48] As mentioned in Chapter 4, military activity now accounts for around a fifth of all human environmental impacts, which is not surprising when one compares the fuel-consumption of modern tanks and fighter aircraft with that of their Second World War equivalents.[49]

The same forces are at work for individual people and firms. Choices become more and more serious and irrevocable as one's world becomes more unequal – and this goes for everyone, from the precarious government minister with his nuclear deterrent, to the precarious worker with her mobile phone. Homelessness can become a real possibility for people who never thought it could touch them. The superficial 'ordinariness' of everyday life conceals a multitude of uniquely different melodramas. Outwardly, we and our neighbors seem to share a 'level playing field'; inwardly, each of us inhabits his or her own private landscape of threats and possibilities, where an otherwise trifling diversion of just a few meters can involve two thousand meters of descent. Preserving one's options, fear of commitment to some particular option, account for all manner of otherwise inexplicable behavior, and even get in the way of survival.

CLOSING THE TECHNOLOGICAL FRONTIER (OR TRYING TO)

The positionality/commodification spiral tightens the screw on the source of innovation itself: self-directed work. It creates conditions that could never produce the technology it has chosen to stake its future on, or even do anything of lasting value with it.

An entrepreneur hands serious hostages to fortune if he depends more than his rivals do on particularly skilled workers, who might go off ill, have babies, die, demand better pay, or go and work for another employer. The smart money is on those entrepreneurs who can eliminate skilled jobs or organize them in such a way that almost anyone can do them – reducing a job to a relatively easily learned list of routine elements and turning the computer into a potentially lethal weapon. As Ursula Huws has pointed out, this vast extension of hierarchical power over the workplace has blurred old distinctions between public- and private-sector work.

In the early days, governments often saw computers as a way of providing new services (like Britain's Giro bank, launched in 1968), but subsequently they deployed them to commodify services that already existed, breaking the link between service and need, and between the person providing a service and the person served. Activities like teaching, service, retail and even care work could be subjected to the same kind of analysis and control that Frederick Winslow Taylor had introduced to car plants in the early 20th century. These functions might then be sold off to private firms, but even if they weren't, a radical change took place in the balance of power within the workplace. Ursula Huws writes:

> Although it is often associated with privatization, this form of commodification does not necessarily involve a change of formal ownership. It nevertheless brings about enormous changes in the nature of the work... Using principles of 'scientific management' fundamentally unaltered since they were developed in the 19th century by the likes of Babbage and Taylor, work processes are analyzed and broken down into standardized units. Once these standardized units have been defined, performance indicators can be identified and standard protocols or quality-control procedures introduced.[50]

Once jobs are broken down in this way into lists of simple specifications, the individuals who do them become in theory interchangeable 'human resources'.

Rule-driven systems removed workers' discretion from tasks like, for example, approving an overdraft or a loan, or a claim for sickness benefit. As discretion disappeared from the workplace, so did the male workers: jobs in high-street banks, for example, became predominantly women's jobs, with commensurately low pay and status.

Rule-driven systems made it easier to introduce new rules,

for example, specifying exactly what items of identification and surety a would-be borrower (or tenant, or job applicant, or visa applicant) must have, before their request could be considered. If the person lacks a necessary piece of paper, a box cannot be ticked, so the loan, visa, tenancy, benefit payment or whatever cannot proceed, no matter how desperate the person's situation and no matter how much sympathy the worker on the other side of the counter feels for them. In this way, harsher rules can be introduced in the safe knowledge that they will be ruthlessly enforced, even without sadists or psychopaths to enforce them.

In the service of unequal power, the computer has become a pervasive presence, and its direction of development has been turned through 180 degrees. It has become a control mechanism, intensifying inequality by eliminating the kind of informal, autonomous work that produces significant innovation.

OTHER ROUTINES ARE POSSIBLE!

Frederick Taylor and his successors have given routinizing a bad name – but routinizing is part and parcel of doing any job well, and when it is the worker who creates the routine it is a source of satisfaction and even pleasure. Craft workers have always devised jigs, stops and other assemblies to eliminate the need for constant, conscious attention, to prevent error and delay, and to assist accuracy and speed. Scientific instruments, tapestries and top-quality pork pies would be impossible without this kind of thing. Craft industries are full of routine procedures that have to be mastered and internalized; once mastered you can talk and think about other things while doing them, and sometimes even sing. And when you control your own routines, working hours are not so bad and sometimes good.

Even today, people like lawyers, writers, entrepreneurs and financiers, and of course the computer programmers (especially the ones who provide the free software that the lawyers, writers, entrepreneurs and financiers would be sunk without) know all about long days that nonetheless are some kind of fun, so the phenomenon cannot be brushed off as a nostalgic invention.

It's a nice irony that while computerization has made it possible to routinize more and more kinds of work, computer programming itself has remained highly resistant to routinizing by employers, so that programmers have been able to keep control of their own techniques and culture. Those who write about programming for other programmers often draw attention to its similarity with older craft traditions.[51]

The craft of programming hasn't survived because programmers are particularly well organized (they are not!) or because employers didn't try to take control. During the late 1970s and 1980s business leaders were told that '4GLS' (fourth-generation languages) and 'CASE tools' (Computer-Assisted Software Engineering) would make it possible for managers to do their own programming and free them from reliance on temperamental and hard-to-manage programmers. Some analysts, like Phil Kraft and Joan Greenbaum, felt certain that programming would become a deskilled, routinized occupation, like so many other professions.[52]

But this didn't quite happen, to a large extent because of capitalist industry's inability to let go of the first architecture that came to hand – the one drafted by John von Neumann in 1946 (see Chapter 1) – and the constant changes that have been needed ever since to overcome inherent inefficiencies, so that the machines and systems people work with continually change in little, messy ways that 'break' previous code. The problem has been compounded by the constant game of push-and-shove whereby software companies seek to 'own' the programming process, by bringing out different proprietary languages and development tools, which also then have to be updated at every touch and turn. How capitalism got itself into this fix is examined later, in Chapter 11.

WHO DECIDES WHAT FOR WHOM

What makes the difference between the satisfying, life-affirming routine of the Edwardian wheelwright or 21st-century geek, and the deadly, mind-and-body crushing routine of the auto-worker, is who has done the routinizing, and for whom. The issue here is autonomy (where decisions are made on one's own say-so) versus 'heteronomy' (where they are made 'in the name of someone else', to translate the term literally). Heteronomous decisions are made by, or in the hope of pleasing, another person, or entity – possibly even a supernatural one, or one so remote that it might as well be supernatural. For the Greek economist and philosopher Cornelius Castoriadis, capitalism and Soviet communism both failed catastrophically, if in different ways, exactly because they were both defined by heteronomy – a principle whose purest expression is cruelty and violence.[53]

Under heteronomy, the worker follows orders or else; no wonder the work is miserable. Autonomous workers, on the other hand, define their own routines, use them as and when they find them useful, and adapt them to the situation without anyone else's say-so. Risk-taking is inherent to the task, but the risks are small ones. David Pye, a designer-craftsperson and teacher at the Royal Academy of Art – and one of the few writers to distinguish between craft and design – called this 'the workmanship of risk': work where one is constantly testing a boundary between what's OK, what's excellent, and what's a mess because you've pushed your luck a bit too far. Computer programming is full of that kind of thing. Pye said:

> all the works of men [sic] which have been most admired since the beginning of our history have been made by the workmanship of risk, the last three or four generations excepted.[54]

Pye was not harking back to some golden age of handicrafts; he envisaged a continuance of the long-established trend in workshops, for workers to use the latest technologies in support of their craft. Exactly the same concern was articulated in a more highly charged political context by people like Howard Rosenbrock and Mike Cooley in the 1970s during worker-led initiatives for 'socially useful production',[55] and in the 1980s within the Greater London Council's Technology Networks.[56] Alas, David Pye's world had little overlap with the workplace struggles taking place around him in Britain in the 1960s and 1970s.

Pye observed that whereas plenty is written and said about the design of the artefacts and buildings around us, their overall design is not as important to our enjoyment of them, as their texture and their surface qualities, which we experience at close range, and which are the results of the work process itself. 'Design begins to fail to control the appearance of the environment at just those ranges at which the environment most impinges on us'.[57]

Risk management is central to autonomous craft, and to the character of its results: parsimonious use of materials, little waste, high functionality, great elegance. The importance of autonomous control of risk is evident, for example, in the work of a virtuoso potter, who can raise the clay on the wheel to a point where in other hands it would collapse in a messy heap, but it also holds true in quite other situations, such as the Finnish education system (mentioned in Chapter 4) where the national inspectorate has been abolished, teachers enjoy great autonomy, and the country achieves the best education results in the world.

Finland may give us the merest foretaste of the sorts of developments we might expect when inequality – and the positional, heteronomous forces that flow from it – go into retreat. Pye suggested that these developments could reach deep into our material world, transforming it utterly – a theme I'll come back to in the final chapter.

1   Thomas Piketty, *Capital in the Twenty-First Century*, Harvard University Press, 2014, p 444.
2   James Boyle, 'The Second Enclosure Movement and the Construction of the Public Domain', The Political Economy of Intellectual Property Rights, 2010, 1: 425-466.
3   Petr Alekseevich Kropotkin, *Mutual Aid, a factor of evolution*, Extending Horizons Books 1955 (originally 1902), p 194.
4   Fintan O'Toole, *White Savage*, Farrar, Straus and Giroux, 2005.
5   Joseph J Ellis, *His Excellency: George Washington*, Vintage, 2004, p 156.
6   Quoted by Simon Fairlie, 'A Short History of Enclosure' in *The Land* magazine, nin.tl/enclosurehistory
7   Robin Jenkins, *The Road to Alto*, Pluto Press, 1979, p 111.
8   Adrian Levy & Cathy Scott-Clark, 'Country for Sale', *The Guardian*, 26 April 2008, nin.tl/Cambodiaforsale
9   Robert C Allen, *The British Industrial Revolution in Global Perspective*, Cambridge University Press, 2009, pp 63-79.

10. John Hammond, Lawrence LeBreton & Barbara Hammond, *The Village Labourer* Vol 1 Guild, 1948, p 83.
11. Kropotkin, op cit, p 182.
12. Karl Marx, *Capital: a critique of political economy*, vol 1, Progress Publishers, 1978 edition, p 71.
13. Alan Story, Colin Darch & Debora Halbert, The Copy/South Dossier Issues in the Economics, Politics, and Ideology of Copyright in the Global South. Copy/South Research Group, 2006, p.27 nin.tl/CopySouth
14. *The Economist*, 17 June 2006, p 33 – report on study by Baruch Lev of the Stern School of Business, New York University.
15. Intellectual Property and the US Economy, US Patent and Trademark Office, March 2012, p viii.
16. Ibid.
17. Xiangming Chen, 'The New Spatial Division of Labor and Commodity Chains in the GSCER', in ed Gary Gereffi & Miguel Korzeniewicz, *Commodity Chains and Global Capitalism*, Praeger 1994.
18. Gereffi & Korzeniewicz, op cit, Chapter 9 by Appelbaum, Smith & Christerson; Doug Miller, *Towards Sustainable Labour Costing in the Global Apparel Industry*, University of Northumbria, 2010.
19. *UNCTAD Trade and Development Report 2002*, quoted by KMG Astill, *Clean up Your Computer – Working Conditions in the Electronics Sector*, CAFOD, Jan 2004, p 8.
20. Thomas Bollyky, 'Why Chemotherapy That Costs $70,000 in the US Costs $2,500 in India', *The Atlantic*, 10 April 2013.
21. Michele Boldrin & David K Levine, *Against Intellectual Monopoly*, Cambridge University Press, 2008.
22. F Bowring, 'Manufacturing scarcity: food biotechnology and the life sciences industry', *Capital & Class*, 27:1, Spring 2003, pp 107-144.
23. Programming Freedom newsletter, Feb 1995, no 11.
24. Bill Gates, 'Challenges and Strategy', memo to Microsoft executive staff, 16 May 1991, nin.tl/Gatesmemo
25. David Graeber, *Debt: the first 5,000 years*, Melville House, 2011.
26. Ann Pettifor, *Just Money: how society can break the despotic power of finance*, Commonwealth Publishing, 2014.
27. John Kenneth Galbraith, *Money, whence it came, where it went*, Houghton Mifflin, 1975.
28. Mary Mellor, *The Future of Money*, Pluto Press, 2010, p 140.
29. Pettifor, op cit, p 33.
30. Mellor, op cit, p 60
31. Ibid, p 66
32. Tobias Preis, 'Econophysics in A Nutshell', *Science & Culture*, 76, no 9/10, 2010, pp 333-7.
33. Mellor, op cit, p 93.
34. Ibid, p 119.
35. Preis, op cit.
36. Michael Black, 'The Sins of Financiers', *Oxford Today* 26, no 2, Trinity Term 2014: 42–44.
37. DA MacKenzie, 'End-of-the-World Trade', *London Review of Books* 30(9), 2008.

38 Danny Dorling, 'The trouble with moving upmarket', *The Guardian*, 18 July 2007.
39 D Dorling, J Rigby & B Wheeler, *Poverty, wealth and place in Britain, 1968 to 2005*, Joseph Rowntree Foundation, 2005, nin.tl/DorlingRigbyWheeler
40 'No more shoddy "Noddy" boxes' says RIBA in call for bigger and better homes', Press release, Royal Institute of British Architects, 4 July 2007, nin.tl/RIBArelease
41 Keith Kintrea, Jon Bannister, Jon Pickering, Maggie Reid & Naofumi Suzuki, 'Young People and Territoriality in British Cities', Joseph Rowntree Foundation, 2008, nin.tl/territoriality
42 Danny Dorling, *All that is solid: the great housing disaster*, Allen Lane, 2014, p 13.
43 Chris Irvine, 'Hospitals and prisons worst offenders in carbon emissions', 1 Jan 2010, nin.tl/CO2prisons
44 Tracy Huling, 'Building a Prison Economy in Rural America', nin.tl/ruralprison Accessed 29 July 2014.
45 Richard G Wilkinson & Kate Pickett, *The Spirit Level*, Bloomsbury 2014.
46 Elisabeth Sköns, 'Military Expenditure', in SIPRI Yearbook 2004, sipri.org/yearbook/2004/10 Accessed 28 March 2014.
47 S Perlo-Freeman, A Fleurant. & ST Wezeman, *Trends in World Military Expenditure 2014*, SIPRI, 2015. Full SIPRI database available at: sipri.org/databases/milex
48 Aude Fleurant, Sam Perlo-Freeman, Pieter D Wezeman, Siemon T Wezeman & Noel Kelly, 'The SIPRI Top 100 arms-producing and military services companies', 2014, nin.tl/SIPRIdatabase
49 See: Patricia Hynes, 'War and the True Tragedy of the Commons', *Truthout*, 28 July 2011, nin.tl/PHynes2011 ; Global Peace Index, visionofhumanity.org ; Campaign Against Arms Trade, caat.org.uk
50 Ursula Huws, *Begging and bragging,* Working Lives Research Institute, London Metropolitan University, 7 June 2006.
51 Pete McBreen, *Software craftsmanship: the new imperative*, Addison-Wesley, 2002; Tom Love, *Object Lessons*, SIGS Books, 1993.
52 P Kraft, *Programmers and Managers,* Heidelberg Science Library, Springer, 1977; and Joan M Greenbaum, *In the Name of Efficiency,* Temple University, 1979.
53 C Castoriadis & DA Curtis, *The Castoriadis Reader*, Blackwell, 1997, Chapter 3.
54 David W Pye, *The Nature and Art of Workmanship*, Cambridge University Press, 1968.
55 Mike Cooley, *Architect or Bee?* Hogarth, 1987; and Mike Cooley, 'From Judgment to Calculation', AI & Society, 21, no 4, 2007, 395–409.
56 Adrian Smith, *Socially Useful Production*, STEPS Working Paper 58, University of Sussex, 2014 Available at steps-centre.org
57 Pye, op cit, p 62.

8

Sales effort: from the automobile to the microchip

In a modern economy, more and more of what looks like productive activity is in fact sales effort. This, plus manufacturers' fear of reliance on skilled workers, has shaped the automobile as we know it, then aircraft. Microchips and microprocessors are shaped by the most extreme sales pressure yet, which spills over into almost every other industry as chip-makers scramble to find new markets for their latest devices before they go obsolete. A whole additional level of sales effort has emerged, aimed at shaping future consumer demand, in the form of the high-tech visionary movement.

In 1966, two Marxist economists, Paul A Baran and Paul M Sweezy, presented an analysis of capitalism's evolution into an era of giant corporations, in which competition had been eclipsed by oligopoly and monopoly, and the problem of 'diminishing returns' by a surprising tendency of surplus to rise. Their book *Monopoly Capital* presents a picture of capitalism caught, like the Sorcerer's Apprentice, between the need to extract bigger and yet bigger profits, and then find ways of investing those profits, so that they generate yet more profits, for which yet more profitable investments must be found, ad infinitum.

The surplus, for example, must be increased in order to support an ever-rising share-price and dividend, so wages must be suppressed, but that imperils demand, which threatens a slump – which is averted, traditionally, by spending on luxury goods and weapons, wars and their aftermaths, and prestige items (cathedrals, corporate headquarters buildings and salaries etc). Baran and Sweezy observed that two further escape-routes had presented themselves since Marx's day: 'epoch-making innovations' and 'the sales effort'.

Epoch-making innovations are ones that cause a wholesale rearrangement of the fabric of life so that everything has to be built all over again, absorbing the problematic surplus. Railways, electrification and automobilization were the examples Baran and Sweezy had in mind. Electronics and computers soon proved to be an even bigger case in point.

Baran and Sweezy's second 'escape route', the sales effort, had already become much bigger than mere advertising. Branding had become a powerful economic force well before the end of the 19th century and some of the world's biggest fortunes were made by the owners of well-known brands attached to essentially minor products: Pears Soap, Coca-Cola, Pepsi-Cola, Colman's mustard-powder, Goddard's silver polish, Oxo stock cubes, Lea and Perrin's sauce. France's wealthiest individual, Liliane Bettencourt, owes her fortune to the L'Oréal brand of inexpensive cosmetics, founded in 1907.

Brands that are well supported by advertising and well defended legally can command far greater, more reliable and more durable premiums (essentially rent) than those attainable by technical improvement. Consequently, branding 'moved up the food chain' in the 20th century, to the extent that (like Nike, mentioned in the last chapter) many manufacturing companies have in practice become brand owners first and foremost, farming out the less lucrative work of production, and even of innovation, to others.

Product design itself became increasingly an extension of the sales effort, rather than an exploration of technical possibilities and human needs. By the mid-20th century this was making it almost (but not quite) impossible to work out what the proper cost of anything would be, if it were produced simply for convenient and comfortable use. At the same time (and reinforcing the tendency) the greater earnings offered by the sales sector were making it a magnet for creative talent that in other circumstances might have sought work in engineering, medicine or the arts.

Recent writers have shown that this complex, self-reinforcing system grows as inequality grows. Thomas Piketty has shown that rent production is more lucrative, and so absorbs more economic effort, in countries and times when extremes of wealth are tolerated. Kate Pickett and Richard Wilkinson have shown that in more unequal countries advertising consumes a higher proportion of GDP than it does in less unequal ones; for example, the US and New Zealand now spend twice as much per head on advertising as Norway and Denmark.[1] Baran and Sweezy focused on the automobile industry, and found mind-boggling discrepancies between what the industry was capable of achieving, in terms of affordable products, and what it was actually producing.

Styling had become such a focus of corporate research effort that vehicle safety was compromised (and erupted into public scandal with the publication of Vance Packard's 1960 bestseller, *The Waste Makers*,[2] and in 1965 by Ralph Nader's *Unsafe at Any Speed*[3]).

Using a study by MIT economists Franklin Fisher, Zvi Grilliches and Karl Kaysen of the costs of styling changes to automobiles between 1949 and 1960,[4] Baran and Sweezy estimated that an average car (then costing around $2,500) could under a purely function-oriented system be produced for around $700, and be more reliable, convenient and durable.

THE ALL-STEEL AUTOMOBILE AS AN ENERGY SUMP

Monopoly Capital was not taken seriously by the economic mainstream, but a few years after it appeared, a fuel crisis triggered a surge of official interest in the impacts of automobile manufacture. In 1973 the State of Illinois commissioned two physicists, R Stephen Berry and Margaret F Fels, to carry out a thermodynamic analysis of automobile production – an analysis of all the energy costs involved, from mining the deposits of iron ore to the showroom.

First, using government manufacturing statistics, they

calculated the amount of energy in kilowatt-hours (kwh) used in manufacturing a typical 1967 automobile, weighing 1.6 tons (most of it steel). This came to 37,275 kwh per vehicle – mainly the cost of turning metal ores into the necessary steel and other metals.

Next, they worked out the 'ideal' energy cost for the same amounts of metals. They looked at the actual chemical changes involved in transforming ores into auto-grade metals, and calculated the known amounts of energy consumed or released by each change (the 'absolute thermodynamic potential change'). This came to just 1,035 kwh per automobile: about three per cent of the amount actually used.

This rather striking difference arises from the fact that industrial metal production is still and may always remain an imperfect art. However, there is clearly plenty of room for improvement. Berry and Fels suggest that it would make a lot of sense to make improving these technologies an 'institutional social goal'. But even so, using 1973 technology and resources, and making optimal use of recycled metal (which needs much less energy to smelt) they reckoned that it was perfectly possible to reduce energy cost by 12,640 kwh (more than a third). But the biggest savings could be made simply by increasing a typical car's lifetime. Extending it from 10 to 30 years would save a further 23,000 kwh: a total saving of 96 per cent.

Since Berry and Fels made their calculations, nothing much has changed in the auto industry apart from the replacement of some mechanical and electro-mechanical components by electronic ones (which cannot be repaired easily or cheaply). And, despite improved recycling methods, even less recycled metal is used in cars; and even less of the metal that is used in them can be recycled (because of new alloying and laser-welding techniques, whose main purpose is to assist the design of eye-catching shapes).[5]

There is one final twist to this extraordinary tale: how cars came to be made of steel at all. Most automobiles were originally made largely of wood, using well-established and highly refined coach-building techniques. Wood is strong, resilient, easy to repair and modify, it does not rust (a major problem for owners until the 1980s, when the electrophoretic method of painting was introduced), and is much lighter than steel. In 1920, according to the UK's Thatcham Motor Industry Research Centre, '85 per cent of the vehicle body was constructed of wood but within six years the wood content had been reduced to 30 per cent, being replaced with steel'.[6] Why? Because paint on metal can be dried in hours, by baking; on wood the process could take days, and hastening it by heating was a dangerous option. The article continues:

> This problem was so prominent that it is suggested that Henry Ford's legendary quote 'you can have any color as long as it is black' was stated in the knowledge that black paint was the quickest drying and would result in an increase in output.

It might be argued that making car bodies from wood must have been more expensive, but apparently this was not so. Set-up costs would have been lower, and there was a huge reservoir of skilled labor from the coach-building and wheelwrighting industries, complete with enormous bodies of knowledge, systems of apprenticeship and so on. We get a sense of the scale and nature of this other world, and how very adaptable it was, from George Sturt's 1923 classic, The Wheelwright's Shop.[7] And to get a sense of the 'look and feel' of the kind of world it could create, go to any stately home and focus just on the visual and tactile qualities of what you find there.

The shift to steel was almost certainly also motivated to some extent by the capitalist's ancient preference for unskilled, unorganized labor, and fear of being reliant on employees' personal skills.

As soon as Ford adopted all-steel, production-line manufacturing, the full force of positional competition came into play. All other manufacturers had to follow suit or lose their markets. As Ford increased the speeds of their production lines, so did everyone else; and everyone, including Ford, had to fall in line again when General Motors introduced annual model changes – a development that had only become possible thanks to wood's displacement by steel, and its ability to be formed, cheaply, into almost any shape, just by changing dies.

This shift was above all else a positional phenomenon, starting with a bold decision by Ford to take control of all the options and decisions involved in manufacture, shifting to new methods and materials so that this could be done. He gambled that the inferior paint finishes would be of less concern to customers than the price and availability he could offer (if it all worked). From this flowed the familiar process of labor-alienation described by just about every book on auto-work since then (of which Paul Stewart and his colleagues' 2009 book *We Sell Our Time No More* is a particularly rich example[8]). Needless to say, there is an utter lack of overlap between that literature and the world of auto magazines, ads and TV programs.

As Berry and Fels's calculations show, the same power shift coerced nature into the process on an unprecedented scale; and we can now recognize this as a positional phenomenon.

Finally, a powerful cultural assumption had been created and then brought into play: that cars are made of steel; those that aren't are odd or old-fashioned (and therefore presumably don't work as well). Practical (and highly positional) considerations also come into play as investment flees technologies based on the older material, insurers demand higher premiums for insuring it, and so on, so that it becomes impossible for an individual or even a substantial group to buck the trend.

CULTURAL 'MATERIALISM'

The assumption that steel is the obvious choice for automobiles could be called a kind of 'materialism': a discriminatory 'ism' of the same family as 'racism' and 'sexism'. It runs deeper than we generally recognize. For example, during the Second World War and the immediate post-War decades, enormous progress was made in the development of new composite materials. One of the pioneers of the new science, JE Gordon, continually referred his students and readers to the properties of the original composite material, wood, even for aircraft, but felt he was swimming against the cultural tide: 'metals are considered "more important" than wood', he wrote. '[It] is hardly considered worthy of serious attention at all'.[9] Yet size for size, wooden aircraft

> weighed less than a 10th of modern hard-skinned machines... In its better forms... the wooden biplane is almost everlasting... they are much longer-lived than motor cars... only the other day (1975) I saw a de Havilland Rapide... flying around very happily; it had probably passed its 40th birthday'.[10]

War is the ultimate positional game, and it played a big part in the transition from wood to metal. The need for speed tells only part of the story – some wood-and-canvas and plywood aircraft were just as fast as aluminum monocoque ones, could fly higher, and their construction required far less energy (aluminum smelting is even now one of the most energy-intensive processes on earth). Wooden-framed aircraft were also much more resilient and simple to repair. The larger reason for their supercession was that construction and maintenance needed more skilled labor, with a much greater understanding of materials: resources that the anxious or ambitious planner or manufacturer could never hold in his own two hands.

Whether or not aluminum monocoque technology is necessary for modern airliners, they might not even have been under consideration had it not been for the Cold War (and, as I mentioned in Chapter 1, the Boeing 707 – developed at public expense as a military transport while Cold War hysteria was at its peak). A further intriguing fact mentioned by Gordon is that their payloads could have been increased threefold, even in the 1970s, by building them from modern composites, but this has not happened – further evidence that when power inequalities are at work, technological development is neither inevitable, nor guided by straightforward goals and rational means. Speed might seem to be an overriding consideration, but aircraft speeds have not increased much since the 1960s. The Anglo-French Concorde demonstrated that supersonic airliners were possible, and it operated for 27 years, but it turned out not to represent a future that the aviation industry as a whole wanted, and it was withdrawn from service after its one and only disaster, in 2003.

Composites have only found 21st-century niches in their more exotic and high-profit forms, such as Kevlar for soldiers' helmets and body armor, a few prestige military aircraft, high-end sports equipment, and a very small number of genuinely transformative products, which are not made by large firms, or even firms at all, such as improved artificial limbs.

Science historian Eric Schatzberg has written about the way metal came to replace wood in aircraft and concludes that 'proponents of metal used rhetoric to link metal with progress and wood with stasis' so that 'a far greater proportion of resources went into improving metal airplanes'.[11] We will find the same 'materialism' at work when we look at how computers took their present form, in Chapter 11.

HOW THE SALES EFFORT SHAPED THE CHIP

The basic building block of almost all the digital electronic devices we use is 'the chip': a tiny, solid-state device made of selectively 'doped' layers of the common element silicon, so that it can be made to conduct or not to conduct electricity a few electrons at a time (hence, it is an 'electronic' device, rather than an *electrical* one which carries vastly heavier current and can give you a shock). The ability of this semiconductor, as it's called, to switch between conducting and not conducting is the current basis of electronic digital computing: zillions of little switches going on and off, passing tiny electrical charges around very fast.

We tend to assume that the silicon chip, and the form of computing that follows from it, were necessary and obvious developments. But this particular technology possibly would not exist, or would not have become anything like as dominant as it has become, without Cold War anxiety to build nuclear missiles to counter a supposed threat to the US from Soviet long-range bombers. The historian Paul Edwards records that, between the 1950s and 1970s, nearly half of the cost of developing integrated circuit technology was paid for by a single missile project, the 'Minute Man'. Up to 1990, up to 80 per cent of all the research carried out on 'artificial intelligence' was funded by the defense research agency ARPA. 'The computerization of society,' says Edwards, 'has essentially been a side-effect of the computerization of war'.[12]

The silicon chip's evolution since then has been shaped to a large degree by a self-reinforcing need to maintain sales. This has produced extraordinary increases in performance, but at considerable cost, and the benefits of that increase are not as clear as we think. The fundamental design principles of chips and the computers they are used in have not changed since John von Neumann wrote out his specification in 1946 (see Chapter 1). Fast though they now are, most computing devices can still only do one tiny thing at a time, so the speed increases need to be considerable to overcome the architecture's limitations. Better alternatives have never been able to make headway because the frantic competition between manufacturers does not permit them to deviate from the path they are on – as we will see in Chapter 11.

MOORE'S SELF-FULFILLING PROPHECY: CHIPS WITH EVERYTHING

The microchip owes a great deal of its success to its role as a commodity in the traditional sense described in Chapter 7: something that can theoretically be turned out in any quantity one wants, as if by turning a handle. If the ideal of commodity production is printing banknotes, microchip production comes surprisingly close to it – not just in the abstract sense that each chip is identical, has a vouched-for provenance, and commands an internationally recognized price that declines over time, but also in the literal sense that chips are produced by processes based on ones that were originally developed for the printing industries: photo-lithography and etching.

The microchip is the first machine that can in principle be produced in the same way as copies of a best-selling novel are produced, in potentially limitless quantities, on demand, for as long as its price holds up in the marketplace. The aim of the game is to cash in while demand lasts, and the industry is shaped by that requirement. As with printing and publishing, the process depends heavily on highly skilled and inherently hard-to-control labor in the design and setting-up stages but, in principle, it is almost entirely free of labor constraints when it comes to production. In principle, all that's needed is able bodies to deliver the goods and haul in the proceeds. But it is also true, as with printing and commodity production generally, that 'in practice' is tantalizingly different from 'in principle'. The quest to solve that discrepancy has turned the industry into a global phenomenon.

Computer-chip manufacture is a high-stakes game: hugely profitable, but with the catch that serious profits are only made on the fastest and latest chips while they are still 'leading edge'.

The industry is said to be driven by 'Moore's Law', which began life as an observation in 1965 by Intel's co-founder Gordon Moore that transistor densities had doubled every two years or thereabouts – a trend that then continued with impressive consistency. An Intel microprocessor of 1972, the 8008, had 3,500 transistors; its successor in 1982, the 80286, had 134,000; in 2000, the last of that particular line of processors, the Pentium 4, had 42 million[13]... and so it continues.

This trend is usually spoken of as if it were a natural law but, as sociologist Donald MacKenzie pointed out in 1996: 'in all the many economic and sociological studies of information technology there is scarcely a single piece of published research... on the determinants of the Moore's Law pattern.'[14] It is, he contends, a 'self-fulfilling prophecy'. The small number of firms with the resources to manufacture chips, but no means of co-ordinating their actions, have been drawn into a kind of bidding war – so that they end up committing astronomical and yet-more astronomical amounts to stay in the game – with relatively little positive effect on the wider world: the same kind of phenomenon as a traffic wave, a stampede or a stock-market bubble.

In 2013 the microprocessor industry was worth $213 billion[15] to the 25 biggest manufacturers (Intel being the biggest; more than twice as big as its nearest rival, Samsung). It is extraordinarily capital intensive and getting more so at an extraordinary rate: a fabrication plant (known in the industry as a 'fab') cost around

$3 billion to build in 2003 and that cost was said to be doubling every four years.[16] By 2005, the most expensive new 'megafabs' were said to be costing around $10 billion.[17] As Nathan Rosenberg has shown in his book *Inside the Black Box*, this makes it ever more important for the manufacturer to make each new device applicable to as many functions as possible. Merely serving a niche market would be financial suicide. 'A firm with a nonstandard component would be forgoing the bulk of the market and would be sacrificing attendant economies of scale'.[18] This means that any and every industry, however apparently non-technical, is likely to be a target for computerization, as we see in the conversion of so many industries from old established technologies to digital.

Hence, devices are integrated into standards-compliant packages – eventually, on a single chip – that can interface with everything they might be required to be used with. The process, known as 'Very Large Scale Integration' (VLSI), has come to dominate computer hardware development. Because of the escalating investment cost, and the very brief time window for recouping those costs before the device becomes 'generic' or obsolete, it is absolutely vital to sell its benefits and possible uses to as many other manufacturers as possible.

In consequence, microprocessors turn up in everything that can conceivably accommodate one. Just two per cent of all microprocessors sold went into personal computers in 2009 and a similar but growing number into mobile phones.[19] From the early 2000s, a new generation of phones, the 'smartphones' (like the iPhone and Blackberry), and then tablet computers 'came to the rescue' of the chip industry because they needed large numbers of the high-profit leading-edge microprocessor chips: each of these devices used around four microprocessors in 2008, against just one or two lower-specification ones for basic or 'feature' phones.[20] The vast majority of microchips are used for 'embedded control' – in washing machines, automobile engine-management systems, avionics, industrial process management, distribution systems, telephone switching systems and the telephones themselves, audio systems and so on.

These devices and systems are steadily being closed off to

user intervention or even repair – as when the various trades involved in coachwork were locked out of the auto industry by the introduction of steel in the 1920s.

The peculiar economics of chip manufacture mean that there is very little profit to be made from older chips (I'll explain more about this in Chapter 9). Firms cannot hope just to make a living; they have to make a killing, consistently, or perish. There is just a brief window of opportunity in which to produce and find markets for the latest high-profit, high-specification chips that go into 'leading edge' (in other words, premium-priced) consumer products. This, as much as public demand for new alternatives to things like 35mm film-based cameras and vinyl long-playing records, is the reason for the spectacular growth of digital photography, digital audio and TV, smartphones and games devices.

To take a specific type of processor, the digital signal-processor (DSP): DSP chips are typically used to convert an analog signal, typically the tiny, fluctuating electric current that a microphone generates from real-world sounds, into a stream of digits that can be manipulated by a computer program (or conversely, from digital into analog form through a loudspeaker). But they can also be used to filter graphic and video images in various ways, not to mention statistical data, seismic data, data from ultrasound scanners; they are even found in the anti-lock braking systems of automobiles, in fax machines, disk drives and DVD players. When DSP chips first appeared, an important market turned out to be 'speak and spell' toys and their current largest application, the mobile phone, was by no means at the front of anyone's mind. Clearly, a successful DSP device had to be one that could slot into a number of these disparate markets as they appeared, with minimal fuss – and that is what Texas Instruments did, thereby capturing the market for DSP.

This tendency or pressure to standardize created 'unprecedented and unanticipated opportunities to use and recombine devices in new ways', says Rosenberg. But it has also meant that devices now *have* to be recombinable to justify the ever-increasing cost and risk of developing them. Development teams need to be able to 'sell' their ideas to management and to the marketing department, showing that they can be incorporated into a great many product-types. They should also have uses within as many different industries as possible, to avoid over-reliance on any particular market sector, which might suffer a downturn or a slump.

Constant, serious sales effort is required at every level. Some of this goes under the name of 'market research' but its aim is not so much to find out what people need, in order to go back to the lab and make something that will meet that need, as to find ways of meeting the manufacturer's need for new markets before their present ones vanish.

DICTATING THE FUTURE

More than 20 years ago, an Oxford sociologist, Steve Woolgar, wrote an amusing article called 'Configuring the User', which showed just how intensely computer-users' wishes were already being stage-managed by experts, supposedly in the interests of finding out and giving them what they want. After spending some weeks with a 'usability' team working on a new microcomputer, he observed that:

> Since the company tends to have better access to the future than the users, it is the company's view which defines users' future requirements.[21]

Woolgar noticed that, as if to add insult to injury, an imaginary entity called 'the user', disempowered in the flesh, had become a handy positional asset in corporate power play. He noticed 'Horror stories about what "users" do to the machines, and conversely a sort of adoption of "the user" as a holy icon or weapon in arguments'.[22] Similar distortions have been working their way up and down – and shaping – the electronics food chain for decades.

Chip manufacturers must sell their products to the 'OEMs' (Original Equipment Manufacturers) and to their customers – equipment firms, government departments and the computer press and opinion-formers that all these people read.

Because of the complexity of the product, the sheer range of possible uses, the tiny time window of profitability, simple diffusion of knowledge through traditional, scholarly and professional networks is hopelessly insufficient, so a whole new sector of 'educational/sales' activity has evolved in parallel with chip development itself. This runs from the provision of Software Development Kits (SDKs) and tools and courses for clients' own developers, to expensive launch-events, seminars and 'webinars', to a new category of what Rosenberg calls 'technological gatekeepers': consultants, journalists, industry analysts, forecasting firms – all of whom 'are crucial to the diffusion of electronic technologies'. They 'may know the domain of ostensible application, but not necessarily the electronic/digital technology that might be applied to it'.

Training has become crucial to winning and keeping customers, and all software and hardware companies devote energy and resources to it. In the professional literature, one reads (and has read for years) how such-and-such a new technology (ActiveX, ASP, C#...) is flawed in certain unfortunate respects, and bloated, but that it will probably succeed because the 'hard-pressed IT manager' knows he can fall back on the lavish support and training offered by Microsoft (or Oracle, or Cisco, as the case may be).

Ordinary computer users are only aware of this to the extent that they depend for technical support on other members of staff who disappear from time to time to attend Microsoft, Oracle or perhaps Cisco-accredited courses.

All of these ancillary activities, which are certainly training in some sense, are also essential parts of the sales process. The information covered is highly specific to the particular manufacturer (which is part of its appeal). Less and less training of this kind takes place in the traditional way, in Further Education or Community colleges, let alone under apprenticeship schemes – which in any case struggle to keep up with the rapid rate of change in job requirements as computer work becomes more and more a matter of learning how to use particular software and hardware packages, and less and less a matter of mastering basic principles – giving rise to the 'learning the same thing over and over' phenomenon social geographer Chris Benner noted in 1999, in connection with 'lifelong learning' (see Chapter 6).

In 2002, Benner published a detailed study of the way this new, constantly changing, learning and employment regime was playing out among computer workers and their employers in Silicon Valley. A major unexpected outcome was the massive expansion of recruitment firms such as Adecco and Manpower: skills were now seen as ephemeral, so employers were less and less willing to engage staff directly (with the expensive long-term commitments that entailed), hiring them instead project by project, and disposing of them again as soon as possible. Benner's study suggests that the expansion of precarious employment throughout the economy in the 1990s, and the rise of outsourcing firms, started in Silicon Valley – to a large extent as a side-effect of Moore's Law, and the logic of sales pressure, as the chief determinant of the course of the electronics industry. Prior to that time, recruitment agencies had been a small sector, providing mainly secretarial 'temping' services.[23]

THE VISIONARY TURN

Sales effort in the high-tech age has become a culture-wide force that embraces areas of the media, politics and academia previously assumed to be quite separate from the advertising industry. As citizens, we have never had the time, information or space we need to discuss and articulate our possible needs as technology advances; instead, visions of possible futures are thrust at us, all apparently worked out in impressive detail and ready to go (and likely to go with us or without us).

High-tech visions, and the people known as 'visionaries' (and even 'evangelists'[24]) who produce them, have become such basic ingredients of the consumer and business media that we do not really think of them as sales effort – yet great effort and expense have been lavished on them by the electronics and computer firms and those around them.

The 'visionary turn' became institutionalized in 1985, with the creation of the MIT Media Lab by Nicholas Negroponte (wealthy and well-connected professor of architecture, brother of President

George W Bush's Director of National Intelligence and co-founder in 1993 of the influential high-tech style magazine Wired) as 'the pre-eminent computer science laboratory for new media and a high-tech playground for investigating the human-computer interface'.[25] Negroponte was able to engage major companies, eminent academics, famous artists and the military in lavishly funded, futuristic projects, ostensibly 'inventing the future' (the subtitle of a book about the Lab[26] written in 1987 by Negroponte's friend Stewart Brand, the founder of the 'hippy bible', The Whole Earth Catalog). Fred Turner's 2006 book, From Counterculture to Cyberculture[27] describes how the Catalog's carefully modulated, cool but unchallenging rebelliousness became a supremely effective sales-script for big business – ultimately epitomized by Apple's ability to present itself as an anti-corporate, 'rebel' organization, long after it had become one of the world's biggest and most ruthlessly monopolistic corporations (strapline: 'Think Different').

The Lab made it its business to produce high-profile, eye-catching 'demos' of its projects, and to produce exciting new words such as 'virtual reality' (3D graphics), 'telepresence', 'personal digital assistant', and so on, and established the genre of 'scientific visionary as showman' familiar from the TED conferences, which were founded by another wealthy and influential architect, Richard Saul Wurman.[28] The subtext is: there is no need to wonder about the future because brilliant minds are working on it, and it will be nice. Everything you could wish for is being taken care of.

The visionary promise is an individualistic, competitive, optimistic one, which comes with an implied warning that nasty things will happen if you don't buy into it: 'think different, or else'. It is hard to reconcile the fact that 'the future' is not quite the bed of roses that was promised, without questioning the massive societal consensus that the visionaries appear to represent. Crazier ideas emerge from the effort to square this circle.

EMBRACING CARNAGE: FAITH IN DISRUPTION

In 1997, a professor at Harvard Business School, Clayton M Christensen, looked at the new precariousness, and the carnage taking place in the name of progress, and persuaded himself that, far from being a disaster, it was a sign of much, much better things to come. A completely new economic era had dawned. Present discomforts merely indicated that people did not yet understand the new world we had entered. The key to this world was to jettison history (because old systems are defunct in the new environment and their rules are bound to mislead us), and become more individualistic and reckless.

The pioneers of this new world, he believed, were the 'small, aggressive startups' which he believed were already devouring the large, complacent organizations of yesteryear by seizing the latest technologies and rushing them to market, in a botched-together state if necessary, rather than waste precious time getting things right. Christensen coined a new name for the phenomenon in his book *The Innovator's Dilemma* – 'disruptive innovation' – and the idea became an instant, wildfire success, turning up in magazines, books, university courses and company reports throughout the economy – and shaping business policy. *Wired* magazine's editor Kevin Kelly rushed out his own bestseller *New Rules for a New Economy* the following year, and many others followed. The idea is still going from strength to strength. This is from a leaked *New York Times* management report, quoted by historian Jill Lepore in an article analyzing the Christensen phenomenon for *The New Yorker* in June 2014:

> *Disruption is a predictable pattern across many industries in which fledgling companies use new technology to offer cheaper and inferior alternatives to products sold by established players (think Toyota taking on Detroit decades ago).*[29]

'Disruption' fitted perfectly with the 'everything has changed', 'don't get left behind!' rhetoric of the high-tech industries and was avidly adopted in Silicon Valley – as George Packer found in 2013 (in his *New Yorker* article on Silicon Valley culture 'Change the World', mentioned in Chapter 1). Facebook's HQ sported posters bearing the motto 'Move fast and break things'.[30]

Lepore's article, 'The Disruption Machine', sketches out the scale of Christensen's success: 'disruption consultants' and 'disruption conferences' and even a 'degree in disruption' at the University of Southern California. She then subjected all of the examples of 'disruptive innovation' in Christensen's book to historical analysis. None of them emerged with a shred of credibility, not even ones from the industry he claimed to know best: computer disk-drive manufacture:

> Christensen argues that the incumbents in the disk-drive industry were regularly destroyed by newcomers. But today, after much consolidation, the divisions that dominate the industry are divisions that led the market in the 1980s.

To take just one example, Christensen had claimed that Seagate, the long-established manufacturer of drives, had fallen prey in the 1980s to disruptive smaller rivals who introduced 3.5-inch hard disks while Seagate was still, complacently, focusing on 5.25-inch models. Yet even as Christensen went to press, in 1997, Seagate was unshaken as the world's largest maker of disk drives and in 2016 remains one of the two largest. Two of the 'disruptive upstarts' that he claimed had toppled Seagate, Micropolis and MiniScribe, had both collapsed and vanished by 1990.

Lepore found similar differences between claim and reality in all the industries Christensen had covered, and that he had ignored the many major examples that ran counter to his thesis. In essence, it boiled down to 'a set of handpicked case studies' that supported his convictions – and proved to be just what an uneasy business elite wanted to hear. 'Disruption by small aggressive startups' is an example of an easily falsifiable idea taking root in a well-insulated elite and then driving policy.

'Disruption theory' has served as a distraction from the wholesale disruption of lives, jobs, workplaces and communities that happens when any new technology comes under the control of a self-interested elite. It is nothing new, but a more advanced version of the phenomenon Baran and Sweezy described in 1967, whereby capitalism fends off implosion by appropriating a new technology, at ever higher human and environmental cost, just as it did with steam power, electricity and the automobile in their days.

The next two chapters describe just two aspects of the 'new economy' in some detail, to gain a stronger idea of the scale of their impact and how it arises in the capitalist environment. First (Chapter 9) we'll look at the microchip industries that underpin the whole sector, then (Chapter 10) at the unexpectedly large impact of the data whose transmission and storage it supports. In both cases, I hope it will be clear that the impact is not caused by the technologies themselves, but by their deployment in the cause of capitalist competition.

1   Richard Wilkinson & Kate Pickett, *The Spirit Level*, Penguin, p 223.
2   Vance Packard, *The Waste Makers*, D McKay Co, 1960.
3   Ralph Nader, *Unsafe at Any Speed*, Grossman, 1965.
4   Franklin M Fisher, Zvi Griliches & Carl Kaysen, 'The Costs of Automobile Model Changes since 1949', *Journal of Political Economy* 70, no 5, 5 Oct 1962, 433–451.
5   Eugene Incerti, Andy Walker, John Purton 'Trends in vehicle body construction and the potential implications for the motor insurance and repair industries', The Motor Insurance Repair Research Centre, Thatcham, June 2005. In International Bodyshop Industry Symposium: Montreux, Switzerland, 2005.
6   Ibid.
7   G Sturt, *The Wheelwright's Shop*, Cambridge University Press, 1930.
8   Paul Stewart et al, *We Sell Our Time No More*, Pluto Press, 2009.
9   JE Gordon, *The New Science of Strong Materials,* Princeton University Press, 1976, p 112.
10 Ibid, pp 164-5.
11 E Schatzberg, *Wings of Wood, Wings of Metal*, Princeton University Press, 1999; quoted by JS Small, *The Analogue Alternative*, Taylor & Francis, 2013, p 13.
12 PN Edwards, *The Closed World*, MIT Press, 1996, pp 64-66.
13 Wikipedia contributors, 'Transistor count', Wikipedia, nin.tl/transistorcount Accessed 9 March 2016.
14 Donald MacKenzie, *Knowing Machines*, MIT Press, 1996.
15 iHS 'Global Semiconductor Market Set for Strongest Growth in Four Years in 2014', Dec 2014, nin.tl/semiconductorgrowth Accessed 9 March 2016.
16 This is according to venture capitalist Andrew Rock in 2003; see 'Rock's Law' – nin.tl/Rockslaw
17 B Lüthje, 'Making Moore's Law Affordable', *Bringing Technology Back In*, Max Planck Institut, Köln, 2006.
18 N Rosenberg, *Inside the Black Box*, Cambridge University Press, 1983, p 183.
19 Michael Barr, *Real men program in C*, Embedded.com 8 Jan 2009 (retrieved 18 June 2010); see also Jim Turley, Embedded Processors

by the Numbers Embedded.com 1999 and The Two Percent Solution Embedded.com 2002.
20 'Mobile: increasing value per handset', ARM annual report 2008, retrieved 29 Aug 2011, nin.tl/ARMmobile
21 Steve Woolgar, 'Configuring the user: the case of usability trials', *The Sociological Review*, 38, S1, pp 58-99.
22 Woolgar, op cit.
23 Chris Benner, *Work in the New Economy*, ed M Castells, Blackwell, 2002.
24 The term 'evangelist' was deployed for marketing purposes at Apple in 1984; Guy Kawasaki, a member of the original Macintosh team, was the first to bear the title. See Robert X Cringely, *Accidental Empires*, Penguin, 1996, pp 217-8.
25 'Nicholas Negroponte', Wikipedia, nin.tl/negroponte Accessed 1 Feb 2016.
26 Stewart Brand, *The Media Lab*; Viking 2007.
27 F Turner *From Counterculture to Cyberculture*, University of Chicago Press, 2006, p 99.
28 'TED (conference)', Wikipedia, nin.tl/TEDWiki Accessed 9 March 2016.
29 Jill Lepore, 'The Disruption Machine: what the gospel of innovation gets wrong', *The New Yorker*, 23 June 2014.
30 G Packer, May 2013, 'Change the World', *The New Yorker*, nin.tl/packerNY Accessed 12 Dec 2013.

9

Technoptimism hits the buffers

Far from being 'weightless', the computer economy is built upon hardware that has a huge impact on the earth. As components shrink in size, more and harder-to-obtain materials, and more energy, are needed to make them, creating mountains of toxic waste, and becoming waste themselves after just a few years. The system is pushing against a fundamental law of physics,, entropy, but this law offers a new, reliable measure of value, which we badly need.

The late 1990s saw a plethora of books proclaiming the arrival of a 'weightless economy'. The promotional copy for one of the first of them, Diane Coyle's *Weightless World* (1998) described it as 'a world where bytes are the only currency and where the goods that shape our lives – global financial transactions, computer code, and cyberspace commerce – literally have no weight'.[1] Other opinion-shaping titles appearing at around the same time included *Understanding the Virtual Organization*,[2] and *The Death of Distance: How the Communications Revolution will Change our Lives*.[3] The general idea was that tendencies such as 'telecommuting' and 'teleconferencing' would greatly reduce human impact.

But in 2002 a study by Eric Williams and colleagues at the United Nations University found that the weightless world was much heavier than expected.[4] Manufacturing a memory chip weighing two grams required 1.7 kilograms of materials and fossil fuels. In a book published the following year, Williams and a colleague, Ruediger Kuehr, calculated that it took about 1.8 tons (1.63 metric tonnes) of materials to make a single desktop computer.[5] These extraordinary figures arise mainly from the extreme levels to which materials have to be refined, in order to produce reliable devices at such tiny scales as the transistors that make up computer chips.

Suddenly, we see that the 'weightless economy' in fact depends on vast amounts of energy, and billions of tonnes of water and other materials, which have to end up either in the local environment or the atmosphere. The problem of toxic waste becomes more intense as the chips get smaller. Other researchers, including Jan Mazurek, Jim Hightower and Elizabeth Grossman,[6] have been writing books and reports since the late 1990s showing how the problem drove chip manufacture out of its original (heavily polluted) homelands – California's 'Silicon Valley' and upstate New York – and into places where disposal was easier: first, the poorer and more sparsely populated states of the southwestern US, and then the Special Economic Zones (SEZs) of Latin America and Asia.

The toxic nature of chip manufacture goes hand in hand with the product's shrinking size: both are inevitable consequences of Moore's 'Law', which was described in the last chapter and is not so much a real law of physics as a consequence of firms' efforts to stay in a highly rivalrous semiconductor industry. A respected industry commentator, Jim Turley, has explained that chip makers are under acute pressure to cover the costs of a new 'fab' before it becomes obsolete, which they can only do by getting more and more chips out of each silicon 'wafer' (the thin slice of hyper-pure silicon from which the chips are made).[7] This means smaller and smaller chips, from bigger and bigger wafers (currently 300mm diameter is standard, but soon 450mm will be the name of the game) – which means higher and higher risks that tiny impurities and other random factors could compromise production.

The chip makers do not pursue smallness for the sheer love of small things, or a desire to relieve the end-users of excess weight. The pressure to go smaller and yet smaller is built into the game they're in. As Turley explained, chips are produced in such high volumes that their price falls rapidly after their first arrival on the market. Every manufacturer wants to cash in on the high initial price before it drops, hence constant pressure for higher densities, bigger wafers, and bigger, better fabs that can deliver them: a self-reinforcing cycle. The ever-higher cost of each new 'fab' has to be recouped in the same short time-frame – three to five years – and there is no breaking the cycle other than to get out of the industry (which many chip-design firms have done by 'going fabless' – but at the cost of surrendering bargaining power to the specialist 'chip foundries', who have deep enough pockets and strong enough nerves to stay in the game).

Chip makers also have to harvest their money quickly because of the need to set up 'second source' agreements with other firms. These help guarantee supplies of the new chip if demand proves greater than their own plant can meet, or if its output suddenly drops (through purity problems, for example). Without second-sourcing, their customers, the electronics manufacturers, might not risk adopting the new chip, but *because* of second-sourcing, the price is doomed to fall: the chip loses its rarity value, becomes 'generic' and the price drops massively.

The only way out of this trap is forward: bigger and more expensive fabs, doing more complicated things, at higher and higher risk, with more and more effort going into protection and hedging against risk ('partnership' deals; joint ventures; elaborate cross-licensing agreements) and more pressure to find techniques that allow the game to continue. There has never been an industry that required so many new techniques to be invented so rapidly, and brought into play, with as little constraint on cost – except perhaps the nuclear industry in its early days.

Like the nuclear industry, it has won special regulatory exemptions, so that firms can introduce potentially dangerous new chemicals into their processes as and when they see fit, without having them cleared by health and environmental protection bodies. The Clinton administration's 1995 'Commonsense initiative' is a prominent example: introduced at the behest of Intel Corporation, this reduced 'the regulatory burden' on the semiconductor industry by allowing chip firms to introduce new chemicals to their processes without prior clearance.[8]

THE TOXIC DEMANDS OF PURITY

Silicon is one of the cheapest and most easily obtained elements on the planet – but only in its oxide form, quartz. Sand is mostly quartz – and that's where chips come from. Separating the silicon from its oxygen consumes a fair amount of energy: this is done by heating sand (quartz particles) with carbon. The resulting silicon can be between 99.0 and 99.5 per cent pure. This is good enough for most industrial processes, including some quite demanding ones like medical silicone for implants and heart valves, but it is nowhere near pure enough for microchips. At their tiny scale, 1.0 per cent or even 0.5 per cent of impurities would render the devices too erratic to be of any use. So impurities have to be reduced all the way down to below 0.0001 per cent: by four orders of magnitude.

This is done in a series of potentially dangerous, energy-intensive reactions involving vast volumes of chlorine and hydrogen (and both of these, also, must be rendered 'hyper-pure' beforehand, using yet more exotic, energy-intensive processes). Then follows the really expensive stage: slicing the silicon into 'wafers', and cleaning and polishing them to nano-scale perfection. This involves an ever-changing cast of toxic gases and other chemicals, and vast amounts of water. Every day in 2002, between two and three million gallons of water went into a typical 'fab', where it, too, was rendered hyper-pure by yet more energy-intensive processes, before being released back into the environment, with its old, harmless-to-humans impurities replaced by toxic new ones: acids used in etching the transistor itself into the silicon, masking fluids. The process involves a continually changing list of different chemicals, which often have to be hazardous because only highly reactive chemicals will do the jobs they are required to do, to the ultra-high tolerances required.

This is a pure and spectacular example of positional competition driving design: a game of 'faster, smaller, more expensive' that every chip manufacturer is obliged to play.

Smaller, faster microprocessors depend more and more on the so-called 'rare earth elements' (REEs). These are highly reactive, metallic elements whose special electrical properties allow the behaviors of glasses and metal alloys to be fine-tuned. Yttrium is used in computer displays, enhances conductivity in metals, and is important for the kinds of high-power light-emitting diode (LED) lights that are now so common. Cerium is essential for the highly polished glass used in the iPad and similar displays. Neodymium multiplies the power of iron magnets – hence, it became essential in the drive for smaller, higher-capacity disk drives, electric motors, speakers and earphones.

None of these elements is, in fact, 'rare'; they are just extremely difficult to separate from the ores in which they are found. First, this is because they occur at low concentrations: according to research carried out by Kiera Butler for the magazine *Mother Jones* in 2012,[9] the best-quality ores contain only between three and nine per cent of the desired element, so vast amounts of rock must be processed to extract a few kilograms. Second, the very properties that make REEs so desirable render extraction incredibly energy intensive: they form intricate and powerful bonds with other elements, so isolating them requires a succession of extremely energy- and chemical-intensive processes. Butler says that a typical extraction plant consumes a continuous 49 megawatts, and 'two Olympic swimming pools' worth of water every day' – plus large amounts of sulphuric and other acids and solvents. The waste is radioactive as well as chemically toxic. Thorium, which is usually present, has a half-life of about 14 billion years (in other words, its radioactivity may outlast the Earth itself – which is a mere 4.5 billion years old).

Consequently, these elements cannot be produced in the quantities that industry needs anywhere near the sorts of places where they are eventually used.

Kiera Butler visited the Malaysian town of Bukit Merah, where a company called Asian Rare Earth (partly owned by Mitsubishi) had been disposing of millions of gallons of radioactive waste into the local waterways for 30 years, causing miscarriages, birth defects, leukemia and other cancers. These effects eventually provoked mass protests which achieved the plant's closure – but neither a proper survey of the health effects nor a proper clean-up operation. All of the drivers hired by one of her interviewees to get rid of the waste had died young. What is more, in 2008 an Australian company, Lynas, gained approval from the Malaysian government to open a huge, new rare-earth refinery in the east-coast town of Kuantan, intended to produce a fifth of global demand. The ores will be shipped thousands of miles from a mine in Australia. Lynas claims this is justified by lower labor, chemical and energy costs but, as Butler pointed out, there were other possible attractions: Malaysia's environmental-protection laws are much less stringent than Australia's; its environmental movement, although it is a strong one, is nowhere near as well resourced or connected; and its government is much more amenable to corporate deal-making, having granted Lynas a 12-year tax holiday on future profits.

Butler explains that pressure for this kind of deal-making has intensified since China (which has the world's main deposits of higher-grade rare earths, and produced about 97 per cent of the world's supply in 2007) cut exports in 2010 by 35 per cent so as to support Chinese manufacturers. This is thought to be one more reason why so much electronics manufacture now takes place in China.

OBSOLESCENCE AND E-WASTE: A TOTAL SYSTEM

Computers and electronics products are classed as 'consumer durables', but their durability is more like that of fashion garments than that of washing machines, cookers or even cars, and much less than that of items like grand pianos, which are still largely products of a craft industry and so not commodities in the full capitalist sense. A Dell computer executive has even been quoted as saying that their products had 'the shelf life of a lettuce', in a report into working conditions in the computer industry by the Catholic charity Cafod in 2004.[10] Like most other reports, it points out that the electronics industry has something else in common with the garment industry: its workforce is overwhelmingly poor, precarious and female, in telling contrast with the lavishly rewarded 'virtual economy' where the computer programmers work, which is just as overwhelmingly male.

It is convenient to blame these injustices on consumers'

allegedly insatiable desire for novelty. But obsolescence, which is beyond their control, drives the process, and the obsolescence, in its turn, is driven as hard as it will go by the men who head the computer firms. There is no fame or fortune to be had from making a computer that will serve its owner for as long, say, as his or her piano, guitar or favorite kitchen knife would – although there is no reason why it should not.

A computer may still work perfectly well, and its owner may type no faster than she did four years ago, but it cannot run the latest version of the software she relies on, which she must buy because everybody else is now using it: a pure positional phenomenon. Or she can no longer get spares for it (because 'nobody' now has machines that need them). Or her new printer won't connect to it (and the old one can't be repaired or upgraded, let alone sold, because there is no second-hand market for it). Or the guy who used to fix the computer has retired and the new guy only knows about the newer machines. Or she's suddenly aware that everyone else is now communicating via Facebook, which is unusably slow on the old machine. Or the BBC has upgraded its web TV viewer, and you need the latest version of the Flash browser plugin if you want to see programs online, and Flash no longer supports machines as old as yours – because nobody uses them any more. And so on: a constantly shifting positional terrain.

Of course, the latest computers are highly desirable things – but their users have other desires (for stability, for example), which go unacknowledged.

As a result of all these positional factors, few people now keep their computers longer than 4.5 years,[11] so that millions of functioning machines are dumped, of which only 12.5 per cent are recycled in any sense – a very broad sense, because this often means shipping them in containers to be broken up, often by children with hammers and bare feet, or just dumped in the deep blue sea. Scandals have been exposed again and again by groups like Greenpeace, the Silicon Valley Toxics Coalition (formed in 1982 when people living in the San Jose area of California discovered they were being poisoned by the supposedly 'clean' new industries springing up around them and were developing cancers), and the

Seattle-based Basel Action Network (formed in 1997 to monitor the UN's Basel Convention on toxic wastes).

The extraction-to-'toxic trash' cycle has been brought to public attention by, for example, the Basel Action Network's report *Exporting Harm* of 2002, and by author-activists like Elizabeth Grossman, in her 2006 book *Toxic Trash*.[12] The scandals have brought about some regulation, especially if they happened close to home, as with the spate of birth defects and cancers due to toxic pollution around Silicon Valley and the Niagara Falls area of upstate New York, in the early 1980s, described by Grossman.[13] The general trend, however, is to 'offshore' the problem. Horrific illegal e-waste recycling operations are discovered from time to time in Africa, China and the Indian subcontinent – wherever there are people desperate enough to do the dangerous scavenging work that the waste-generating economies cannot afford to do, not to mention the brutal work regimes that destroy workers' health, fertility, babies and often their environments as well, for pitifully low pay, or even no pay at all.

I will not dwell on this side of the 'weightless economy', which is described with the care it deserves in a number of places, including Elizabeth Grossman's book, Jan Mazurek's *Making Microchips*, Jim Hightower and colleagues' *Challenging the Chip*, and Nick Dyer-Witheford and Greg de Peuter's *Games of Empire* (which describes the extraordinary exploitation that goes on at every level in the computer games industry).[14] Hsiao-Hung Pai's *Scattered Sand* is based on interviews she travelled thousands of miles to obtain, with workers in some of China's harshest industrial zones, whose lives prove to be every bit as extraordinary and heroic as those of the corporate leaders whose fortunes they have helped to build.[15]

DISPLACING THE PROBLEM TO AFRICA

Elisabeth Grossman notes that Europe has taken stronger measures than the US against toxics – but one wonders whether this merely reflects Europe's lack of a substantial computer industry of its own to match the lobbying power of that in the US. At any rate, in a globalized economy, virtuous national or regional regulations can simply displace and even intensify the problem.

The European Union's Restriction of the use of Certain Hazardous Substances (RoHS) directive of 2002 prohibited use of some of the more toxic materials, such as lead solder. This was achieved after campaigns that lasted years against very effective rearguard actions from the various industries that were affected. By the time it went though, manufacturers had already discovered satisfactory alternative tin-based soldering methods, which resulted in a tin rush in eastern provinces of the Democratic Republic of the Congo (DRC). This was a convenient source as there was little government there at the time and plenty of cash-hungry soldiers and militias to organize extraction (Hutu militias had made the eastern DRC their base following their expulsion from nearby Rwanda after the 1994 genocide). UK Channel 4's reporter Jonathan Miller visited the Bisie mine, the main source of the tin ore, cassiterite, in 2005. A trader told him:

> The miners work for nothing; the soldiers always steal everything. They even come to shoot people down the mineshafts. Yes, not long ago they shot someone. They force the miners to give them everything and they threaten to shoot anyone who argues. They're always ready to shoot. We are really penalized. We earn nothing. But we pay a lot. The soldiers – they are all around us here, but they are in civilian clothes.[16]

Miller wrote: 'This stuff is mined and portered by people who are like cannon fodder for our industries. Five armies have battled for control of Bisie mine in just five years. But still we buy it.'

This came on the back of an earlier scandal about the 'coltan rush' of 2000, which intensified and expanded the Congo's 'Great War' (1998-2004) into hitherto unaffected parts of eastern DRC. Coltan (Columbite-Tantalite) is one of the ores of the metal tantalum, whose value had risen with the demand for better and smaller capacitors: the devices that store electrical charge in electronic devices. Tantalum capacitors are a major reason why modern electronic devices can be as small and as portable as they now are. Development scholar and writer Michael Nest explains in his 2011 book, *Coltan*, that Congolese coltan is a relatively minor source of the metal, but it is a convenient one: in some places the river mud is entirely coltan and in 2000 there was no shortage of willing and unscrupulous agents to organize extraction.[17] The world's main sources of tantalum, by contrast, are two huge, open-cast mines – one in Australia, the other in Canada – which are highly mechanized and cannot easily be adapted to meet the wilder fluctuations in demand. Informal sources are important to all industries, to smooth out the peaks of demand without incurring any further cost when the peak has passed.

During 2000, large stockholders and refiners had become concerned that they might not have sufficient tantalum should there be a really huge surge in Christmas demand for second-generation mobile phones and the new gaming devices, and placed additional orders for ore as a precaution. This caused prices to spike, from $30–$40 per pound at the start of the year, to up to $300 in September. In Congo, this gave some farmers (many of whom had been unable to plant or harvest crops because of the war) a chance to make badly needed cash; but then militias and cash-strapped military units operating from the Great Lakes region, and their backers, moved in and fought for control, making a substantial contribution to the Congo War's estimated five-million death toll.

As Michael Nest explains, nobody gained anything from this mayhem – which was, however, an entirely natural result of large corporations protecting themselves against financial embarrassment, as required by their legal duty to shareholders. As it happened, the stockholders' worst fears did not materialize; demand for mobile phones and PlayStation2s remained manageable, and by October 2001 the price of tantalum ores had fallen back to its early-2000 level.

WE NEED A NEW WAY TO MEASURE VALUE

The above is just part of a very big picture indeed, of a positionalized, and therefore intrinsically high-impact economic system, which is thoroughly globalized. It is a system shot through with the most radical injustice – inflicted almost in its sleep by nothing nastier than scrupulous cost accountancy and fiduciary diligence, within a framework of national and international law constructed on the premise of strong money and property rights.

Money pretends to offer a reliable measure of the value of things but clearly does not. Is an objective measure of value and harm even possible? The environmental impacts, which happen in tandem with the human ones, can at least be quantified in hard numbers, using the concept of 'entropy' (in this context, the degree to which materials become irretrievably mixed together, so that more work is needed to sort them out and make them usable). Timothy Gutowski, one of those developing this method, draws attention to the steady decline in the rate at which materials are recycled, even as they become harder to obtain:

> Overall, the trends show an apparent remarkable reduction in the recyclability of products that is due primarily to the greater material mixing. Given the rather significant resources devoted to developing complex material mixtures for products compared with the rather modest resources focused on how to recapture these materials, it appears that there is reason for concern.[18]

Gavin Bridge, mentioned earlier in connection with the 'Jevons Paradox', says that:

> The variety of elements on electronic circuit boards, for example, has increased from 11 in the 1980s to potentially over 60 in the 2000s as manufacturers have sought to boost product value and performance... Even mature technologies like telephones can use over 40 different materials: contemporary mobile phones, for example, can contain 17 different metals, only 4 of which – gold, silver, palladium, copper – are currently recycled.[19]

To put the above in context, I was once told by a former employee of the UK's General Post Office (GPO) that their old, Type 332, rotary-dial telephone – featured in Alfred Hitchcock's *Dial M for Murder* and innumerable other British films between the mid-1930s and the 1960s – contained just five different materials, and was 100-per-cent recyclable. It was also not a commodity: it belonged to the GPO and remained its responsibility throughout its life.

Figures published by World Resources International (WRI) have shown a more than fourfold increase since 1960 in the world consumption of four basic metals: nickel, chromium, zinc and copper, with a sudden increase in usage of copper between 1995 and 2000 (from around two million tonnes per year to around five million tonnes). For most materials, this is despite the fact that they have become more difficult and need more energy to extract, because the 'low-hanging fruit' – the richest, most accessible deposits – were used up long ago. And, as the extraction rate mounts, more and more of what's extracted is going straight back into the soil, rivers, seas and atmosphere as waste. WRI reported in 2000 that 'Between half and two-thirds of material inputs to industrial economies returned to the environment within a year'.[20]

The amounts of metals produced have increased dramatically, even as major deposits are exhausted and the quality of remaining ores diminishes. This necessarily implies unprecedented levels of pollution and waste, plus greater oppression in the areas where production takes place, and louder justificatory music in the places where the end products are consumed.

The drive to keep costs down mandates tighter controls on labor – labor is one of the very few factors in the production equation from which managers and entrepreneurs can squeeze extra profit when costs rise or prices fall. Gavin Bridge remarks on the 'stark contrasts... between the hyper-mobility of exported resources and the geographical and social immobility of many people living and working in the same space'.[21]

ENTROPY: MEASURING WHAT'S POSSIBLE

Conventional economics doesn't capture all of the economy's costs and, according to its critics, it isn't meant to. The Harvard economist Stephen Marglin argued in his 2008 book *The Dismal Science* that mainstream economics has always been the servant of powerful interests, which naturally want reassurance that what they are doing is for the best, and don't wish to hear otherwise. Mainstream economics, he writes, 'has been shaped by an agenda focused on showing that markets are good for people rather than on discovering how markets actually work.'[22] So economists still maintain that money (as defined by the banking and finance community) is the best measure of the human value of economic activities – even though we see more and more examples every day of environments and lives ruined by the pursuit of monetary profit.

Value is certainly a tricky thing to measure, but ultimately everything has material cost, whether a monetary price is attached to it or not. Eric Williams used a 'life-cycle analysis' approach to work out what microprocessors really cost in terms of materials and energy (and the materials used to produce the energy). This involves tracking down those very things that a capitalist firm must externalize to survive, exposing much larger externalities and knock-on costs than anyone suspected, and producing a bill of costs that corresponds closely with the observed human and environmental impacts.

For example, the full life-cycle cost of the LED lights referred to by 'green growth' enthusiasts like Chris Huhne (mentioned in Chapter 5) is far greater than it seems because so much energy is used in their manufacture. They are semiconductor devices, made in the same high-energy fabrication plants as computer chips. Even so, they could in principle have only about 25 per cent of the lifetime energy cost of old-fashioned incandescent lighting. However, this is without considering why they are manufactured in Special Economic Zones in the Global South rather than in the English home counties or the Berkshire hills of Massachusetts.

LEDs (and microchips) are manufactured in these distant places because it became too costly and too problematic to carry on making semiconductors in more prosperous places. In the SEZs, and in poorer places generally, costs do not have to be accounted for so rigorously, which implies further external costs that can take some tracking down. In addition to the costs that go with semiconductor manufacture, substantial but hard-to-measure costs are certainly incurred in the factories where the LEDs are assembled into saleable products like torches and car tail-lights. Some costs will be almost indecent to quantify in money terms: illness, early death, birth defects and so on. Finally, because the LEDs have been made so cheaply that quite powerful ones can be given away as promotional gifts, they end up being used far more than the previous generation of lighting. So even though an individual LED light may still use less energy over its lifetime than its predecessor, the cumulative effect is an increase on energy used for lighting, which will show up on the 'environmental bottom line' no matter how hard firms work to hide the impacts that come in between. A number of reports and conference papers have tackled this complicated cost-accountancy problem.[23]

MAXWELL'S DEMON: THE SPOILER IN THE GREEN GROWTH DREAM

An additional way of getting a handle on the problem is to recognize the fundamental laws of physics that are involved.

The key here is the Second Law of Thermodynamics (also known as 'The Entropy Law' and 'The Universal Law'). This surprisingly recent law (1824) states what ought to be obvious: in effect, that you cannot get something for nothing, or have your cake and eat it. The Second Law is like the 'bad news' part of an 'I've got good news and I've got bad news' joke: whereas the First Law says energy cannot be destroyed, whatever you do to it, the Second Law says that, while that may be true, the energy may be no earthly use to you after you've put it to use – so use it carefully. Perpetual-motion machines cannot exist in our universe.

As the Scottish physicist James Clerk Maxwell, put it: 'The second law of thermodynamics has the same degree of truth as the statement that if you throw a tumblerful of water into the sea, you cannot get the same tumblerful of water out again'.[24]

The example that the Parisian engineer Sadi Carnot had in mind when he defined the Law was the heat that goes into a steam engine's boiler: it may still be there after it's passed through the engine's pistons, wafting around in the atmosphere, and may linger for a while in the locomotive's metalwork, but you try doing anything useful with it! If you want to perform the same trick again you must get some more coal. The energy *can* be retrieved by clever engineering (as James Watt had realized not long before, and as power stations do with their enormous cooling towers) but it all takes extra work – which also consumes energy, and dissipates materials into the environment in the same way that the engine's steam melts into the atmosphere. It has to be done with care and consideration.

Maxwell proved in the 1860s that gases (including air and steam) are composed of molecules bouncing around at different speeds, and wondered whether there was a way of using this insight to prove the Second Law, and rule out the possibility of perpetual-motion machines once and for all.

In 1871, in a letter to a friend, Maxwell described a small, imaginary being 'whose faculties are so sharpened that he can follow every molecule in its course'. This imaginary being was enclosed in a box full of air: a collection of randomly moving molecules of nitrogen, oxygen and so on, warmer ones moving quickly, cooler ones moving slowly. The box was divided by a wall in which there was a trap door. Using the trap-door, Maxwell's creature would let the fast-moving, hot molecules go into one compartment, slower, cold ones into the other. Eventually this would turn the box into something like the piston cylinder in a steam locomotive: hot (fast-moving) gas, creating pressure on one side; cold, slow-moving gas with lower pressure on the other: the basic requirement for a machine capable of work. Maxwell's question was: does reality allow this, and if not, why not?

This creature soon became known as 'Maxwell's Demon', and it remained a well-loved riddle of physics until, in 1929, Einstein's friend Leo Szilard worked out why the demon would not be able to do the job. His solution hinged on the new understanding of information (previously assumed to be something immaterial, outside physics). Szilard realized that the demon needs at least to be able to recognize the different molecules before it can do its work: that is, it needs information, and this can't be left out of the calculation. And all information, apparently, comes at a price, even the modest information the demon needs. When the sums are done, and even allowing for the demon to be infinitesimally small and entirely without any physical needs of its own, the cost of the information it needs to identify the fast and slow molecules just outbalances the value of any energy it harvests.[25]

PUNCTURING THE WEIGHTLESS ECONOMISTS

Maxwell's Demon is, therefore, a challenge and a warning to capital, and ought to have introduced caution into the idea of a 'frictionless', 'weightless' information economy. Yet the information economy still operates on the old assumption that information costs nothing. Now we are paying the price, as we will see in Chapter 10.

Even as the 'weightless economy' was preparing for take-off, in 1990, the Hungarian information theorist Tom Stonier was teasing out the implications of Szilard's proof. In a fascinating but little-read book, *Information and the Internal Structure of the Universe*, Stonier showed that information has a physical reality – and that structure and information are counterparts of each other.[26]

When structure (for example, the latticework of an ice crystal) is lost through melting, so is all the information implicit in that structure, which is not just something the human aesthetic sense has projected onto it: it is real, even if it is small. Stonier showed that one joule of energy (a tiny amount) corresponds to 10^{23} bits of information: a 1 followed by 23 zeros, or 100 zettabytes, an astronomical number. The fact that so much information can be had for so little energy sounds like good news for the weightless economists. But first, that is without the cost of any technology for handling the information, and that could be exponentially greater. And second, the converse is also true: even a tiny expenditure of energy necessarily implies the loss of an astronomical amount of information and structure, and it will take exponentially more information (involving yet more work and entropy) to put that structure back together.

This is not what the proponents of weightlessness want to hear, but it ought to give pause to those, like the very rich and influential Matt Ridley, who argue that waste is nothing to worry about because all the valuable materials we have wasted are still there, in the slurry-tanks and spoil heaps and waterways, ready to be harvested by clever new technologies. But all new technologies cost energy, structure and information, no matter how ingenious they seem. Ridley writes (in a piece for the Washington Post, entitled 'Why most resources don't run out'):

> Take phosphorus, an element vital to agricultural fertility. The richest phosphate mines, such as on the island of Nauru in the South Pacific, are all but exhausted. Does that mean the world is running out? No: There are extensive lower-grade deposits, and if we get desperate, all the phosphorus atoms put into the ground over past centuries still exist, especially in the mud of estuaries. It's just a matter of concentrating them again.[27]

I love that 'just'. And who is 'we'? It is of course conceivable that there might be some micro-organism somewhere that will do this job for 'us' but somebody will have to find it (and even Maxwell's Demon would need some resources to do that) and the possibility of knock-on effects should not be ruled out. The overwhelming probability is that it would be exponentially more costly than the original practice of digging guano.

Well-structured things are said (by the Second Law) to have low entropy; disrupted things have high entropy, and high-entropy things are generally useless. The tendency of phenomena in the greater universe is towards greater entropy, as the heat and light of the stars dissipates. At some point in the future the universe faces 'heat death': the same temperature everywhere, and physically uniform in all directions. There is no need to expedite this process, in one of the possibly very rare parts of the universe where entropy has gone into reverse – but that is what the dynamic of capitalism does, and what its ideologies promote.

Low entropy has to be created by work – which for humans means also creating heat, and therefore overall entropy, so we do it with care and respect. The extremely low entropy represented by an ingot of pure silicon demands the greatest respect.

The real source of low entropy is life itself. Life, said the Austrian physicist Erwin Schrödinger in 1943, is a system for locally reversing the Entropy Law: for transforming simple chemicals and water into complex and ever more complex, highly structured organisms, and even complex geologies derived from those organisms. He called the process 'negative entropy'; others shortened it to 'negentropy', which is the term people now use.[28]

Life can do this because the earth is an 'open system' – open to assistance from the sun, a tiny fraction of whose incoming energy (less than 0.1 per cent[29]) is harnessed by plants, to create whole landscapes of order and complexity out of simple minerals and sunlight.

Evolution, says Schrödinger, is the result of nature's progressive refinement of its negentropic techniques, creating more and yet more complex and self-sustaining order through ever-more parsimonious use of the same basic resources. Humanity and its technologies ought to be neither more nor less than a continuation of this process – as Brian Arthur's definition of technology implies. And indeed most human communities for most of our history have, by and large, been negentropic. When humans settle in and develop a new habitat, they can and usually do increase the amount of order that it contains, building complex plant communities for their own uses, in places where little grew before. The results are just as wonderful as when lichen and mosses colonize stark, volcanic landscapes, bringing richness where previously there was only sameness. This was the situation in Ladakh, as described by Helena Norberg-Hodge (Chapter 4); also in millions of back gardens and allotments: oases of diversity and detail, contrasting sharply with everything around them.

All of this breaks down when entrenched inequality and hierarchical societies make their appearance. All the careful prior work of nature and humanity is suddenly fuel for quick self-aggrandizement, and its power to do this increases with the efficiency of the available technology, so that the more parsimonious the technology, the worse the consequences, culminating in the situation we now face.

THE BLEEDING EDGE

Timothy Gutowski and his colleagues (2009) write that:

the intensity of materials and energy used per unit of mass of material processed (measured either as specific energy or exergy) has increased by at least six orders of magnitude over the past several decades. The increase of material/energy intensity use has been primarily a consequence of the introduction of new manufacturing processes, rather than changes in traditional technologies. This phenomenon has been driven by the desire for precise small-scale devices and product features and enabled by stable and declining material and energy prices over this period.[30]

So the alleged super-efficiency of electronic devices is only obtained at a huge and escalating physical cost; and (as we will see shortly) is largely propaganda: the now-dominant type of computer has been described with good reason as 'the least efficient machine that humans have ever built'.[31]

1. Diane Coyle, *The Weightless World*, MIT Press, 1998.
2. Bob Norton & Cathy Smith, *Understanding the Virtual Organization*, Barrons Educational, 1998. For a lucid, early critique of this wave of books see Ursula Huws, Material World, Socialist Register, 1999, nin.tl/Huws1999
3. Frances Cairncross, *The Death of Distance: How the Communications Revolution will Change our Lives*, Harvard Business School Press, Boston, 1997.
4. Eric D Williams, Robert U Ayres & Miriam Heller, 'The 1.7 Kilogram Microchip', Environmental Science & Technology 36 (24),2002, pp 5504–5510, nin.tl/microchipweight
5. Kuehr & Williams, 2003, cited in D Pellow, D Sonnenfeld & T Smith, Challenging the Chip, Temple University Press, 2006, p 205.
6. J Mazurek, Making Microchips, MIT Press, 1999; Ted Smith, David Allan Sonnenfeld & David N Pellow, Challenging the Chip, Temple University Press, 2006, nin.tl/ChallengingtheChip ; E Grossman, High Tech Trash, Island Press, 2006.
7. J Turley, 'Semiconductors on a Train', *Embedded Systems Design*, 2 Aug 2006, nin.tl/semiconductorstrain
8. Mazurek, op cit, p 7.
9. Kiera Butler, 'Your Smartphone's Dirty, Radioactive Secret', *Mother Jones*, Dec 2012, nin.tl/MJsmartphonesecret Accessed 6 April 2013.
10. KMG Astill, Clean up your computer, Cafod, Jan 2004, nin.tl/cleanupcomputer
11. Antony Leather, 'Most Computers Replaced after 4.5 Years', Bit-Tech, 12

May 2011, nin.tl/bit-techreplace Accessed 16 Feb 2014.
12 Grossman, op cit; J Puckett & TC Smith, 'Exporting harm', Basel Action Network, 2002.
13 Grossman, op cit, p 79.
14 Grossman, op cit; Mazurek, op cit; Ted Smith et al, op cit; N Dyer-Witheford & GD De Peuter, *Games of Empire*, University of Minnesota Press, 2009; Astill, op cit.
15 Hsiao-Hung Pai, *Scattered Sand*, Verso, 2012.
16 J Miller, 'Congo's Tin Soldiers', 30 Jun 2005, nin.tl/congotin
17 Michael Wallace Nest, *Coltan*, Polity, 2011.
18 Bhavik R Bakshi & Timothy G Gutowski, *Thermodynamics and the Destruction of Resources*, Cambridge University Press, 2011, p 125.
19 G Bridge, 'Material Worlds: Natural Resources, Resource Geography and the Material Economy',. *Geography Compass*, 3(3), 2009, pp.1217-1244.
20 Ibid.
21 Ibid.
22 SA Marglin, *The Dismal Science*, Harvard University Press, 2008, p 3.
23 Navigant Consulting, Life-Cycle Assessment of Energy and Environmental Impacts of LED Lighting Products Part I, US Department of Energy, Feb 2012. See also Timothy G. Gutowski, 'Manufacturing and the Science of Sustainability', Massachusetts Institute of Technology, 2011.
24 Harvey S Leff & Andrew F Rex, *Maxwell's Demon: Entropy, Information, Computing*, Adam Hilger, 1990, p 39.
25 Leo Szilard, 'On the Decrease of Entropy in a Thermodynamic System by the Intervention of Intelligent Beings', 1929. Published as Chapter 3.1 of Leff and Rex, 1990, op cit.
26 T Stonier, *Information and the Internal Structure of the Universe*, Springer-Verlag, 1990.
27 Matt Ridley, 'Why Most Resources Don't Run out', 30 April 2014, nin.tl/Ridleyresources
28 In his 1943 book *What is Life?* Schrödinger called it 'negative entropy'; Leon Brillouin shortened it to 'negentropy' in his 1949 essay 'Life, thermodynamics and cybernetics', published as Chapter 2.5 of Leff and Rex 1990, op cit.
29 The total solar energy absorbed by Earth's atmosphere, oceans and land masses is approximately 3,850,000 exajoules (EJ) per year. In 2002, this was more energy in one hour than the world used in one year. Photosynthesis captures approximately 3,000 EJ per year in biomass.
30 TG Gutowski, MS Branham, Jeffrey B Dahmus, AJ Jones, A Thiriez & DP Sekulic, 'Thermodynamic Analysis of Resources Used in Manufacturing Processes', *Environmental Science and Technology* 43, no 5, 2009, pp 1584-1590.
31 G Dyson, *Turing's Cathedral: The Origins of the Digital Universe*, Knopf Doubleday, 2012.

10

The data explosion: how the cloud became a juggernaut

During the late 1990s and early 2000s, advertising and e-commerce changed the Web from the lightweight information medium conceived by Tim Berners-Lee into a heavyweight sales engine. This had devastating consequences, not only for the bricks-and-mortar economy, but also for global CO_2 emissions, as competition drove firms out of the high streets, into 'the cloud', which, as early as 2007, was consuming more electricity than most entire countries do.

The Web was conceived as a textual medium, and text itself makes modest environmental demands: the complete works of Shakespeare use only five megabytes of computer memory, or just one megabyte when compressed. Michael Lesk, an expert on digital libraries, once estimated that even if one were able to store every word one had heard in one's lifetime, including all of the words heard on TV, movies and radio, it would add up to about 50 megabytes. Adding in all one's reading (assuming this at 354 hours a year) 'in 70 years of life you would be exposed to around six gigabytes of Ascii'. At the time he wrote his paper, the late 1990s, the content of all the books and newspapers published in the world came to perhaps one terabyte per year[1] – a large amount of computer storage at the time but now fairly common; in January 2016, worth about $70.

To put these numbers in perspective, a terabyte is a thousand gigabytes, a gigabyte is a thousand megabytes, and a megabyte is a thousand kilobytes or 'k', and so on. (In the 1980s, a home computer might have 'a massive 64k' of memory; numbers like 64, 128, 256 and so on are common because they are powers of the number 2, which is the basis of a binary computer universe). Table 2 sets them out.

Table 2 Measures of computer memory

1 byte	The amount of physical memory your particular computer uses to store one text character. Until the mid-1990s, this was generally 8 'bits', offering 256 different characters per byte. Nowadays a byte is usually 32 or 64 bits.
1 kilobyte	1,000 bytes
1 megabyte	1,000 kilobytes
1 gigabyte	1,000 megabytes
1 terabyte	1,000 gigabytes
1 petabyte	1,000 terabytes
1 exabyte	1,000 petabytes
1 zettabyte	1,000 exabytes
1 yottabyte	1,000 zettabytes

Because the amount of actual electricity used in electronic devices like these is so minute, there has been a strong tendency, naturally encouraged by the businesses that sell them, to dismiss the idea that electronic data has any serious physical reality. This became such an orthodoxy that respectable economists fell for it, fuelling enthusiasm for 'the weightless economy' described earlier.

But in 2010, *The Economist* drew attention to a 2008 report on the extraordinary amount of digital information flowing into US households: 3.6 zettabytes (or 34 gigabytes per person) per day. It says that:

The biggest data hogs were video games and television. In terms of bytes, written words are insignificant, amounting to less than 0.1 per cent of the total. However, the amount of reading people do, previously in decline because of television, has almost tripled since 1980, thanks to all that text on the internet. [2]

The study also found that 'only five per cent of the information that is created is "structured", meaning that it comes in a standard

THE BLEEDING EDGE

format of words or numbers... the rest are things like photos and phone calls.'

Then, in 2012, Greenpeace discovered that the prize exhibit of the 'weightless economy', the internet, had a bigger carbon footprint than most countries did:

> The combined electricity demand of the internet/cloud (data centers and telecommunications network) globally in 2007 was approximately 623bn kWh (kilowatt hours). If the cloud were a country, it would have the fifth largest electricity demand in the world.[3]

For comparison, the entire population of India used 568bn kWh – 55bn kWh less than the internet. Germany used only 547bn kWh, Canada 536bn kWh, France 447bn kWh, Brazil 404bn kWh, and the UK 345bn kWh. And the internet's impact is rising at a furious rate, with a predicted 50-fold increase in the amount of digital information by 2020, and half a trillion dollars' worth of new datacenters built just in the year 2012-13.

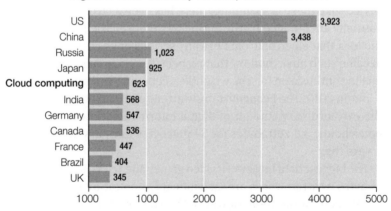

Figure 2 2007 electricity consumption in billion kwH

Reproduced by permission from *Make IT Green: Cloud computing and its contribution to climate change*, Greenpeace, 2010, nin.tl/Greenpeaceoncloud

The main culprit here is 'cloud computing', which gradually trickled into public awareness with services like Gmail (launched in 2004), where your email is stored permanently 'in cyberspace' (in fact, on a hard disk somewhere, in one of Google's data centers: large, air-conditioned sheds that are filled with networked computers known as 'servers' because they 'serve up' data on request). Mail is usually viewed through a Web browser, and not deleted, so that you can check your email from anywhere that has an internet connection and a browser, and can use a number of different devices, perhaps an office computer, a hand-held device, and a desktop machine at home.

Your emails are also, of course, vulnerable to surveillance by government agencies and perhaps others, as the US National Security Agency (NSA) contractor Edward Snowden revealed in 2013.

Hitherto, it had been normal for users to collect their email from their email-providers' servers using PC applications like Thunderbird, Eudora or Apple Mail, which then usually deleted the mail from the server to free up storage space. Under the new regime, all of your email stays on the server, unless you choose to delete it – but you are urged not to, because storage is cheap and getting cheaper, and you never know – you might need it some day.

At its launch in 2004, Gmail offered each user a whole gigabyte (one billion bytes, or characters) of email storage on its free accounts – subsequently doubled and then doubled again. At that time, free email services, such as Hotmail, normally only offered five or ten megabytes (a million bytes) of free storage space for emails, which had seemed generous, but the growing practice of attaching digital photographs to emails, and then video clips, was already making megabyte-sized storage seem restricted.

Google could offer big amounts of storage because of its novel data center strategy: large numbers of them, sited near to centers of population in order to deliver search-results into users' browsers in less time than rivals (with the tiny amount of time taken shown at the head of the page of results). Delays in response were minimized by 'high redundancy' (the sheer quantity of low-cost storage) plus clever load-balancing techniques (jealously guarded commercial secrets, but ultimately derived from the techniques developed during and just after the Second World War for keeping populations fed and armies supplied – as we will see later, in Chapter 12).

Gmail was part of Google's 'win-win-win' strategy for earning money from placing discreet but well-targeted (and therefore premium-priced) advertising into people's web browsers with their search results: a 'win' for the user, who theoretically only gets ads that are of interest; a win for the advertiser, because money isn't wasted on advertising to people who aren't interested; and a financial win for Google. In 2011, most of Google's $37.9 billion revenues came from advertising, making it the world's largest advertising company.

Ads are selected for each user on the basis of a limited amount of information from their emails, and information about their interests, gleaned from the kinds of Google searches they have made (their search history). For example, I use an Apple Macintosh computer, and often search for solutions to little problems that come up; when I did a Google search to try to establish just how big a problem file-duplication is on the internet, Google gave me lots of helpful links to articles about how to get rid of duplicate files on a Mac. At first, this kind of thing raised a storm of hostile comment that Google was abusing personal data, but by and large Google seemed to uphold their claim that they 'don't do evil' – at least, not on that particular issue.

Following Google's success, other big companies, like Microsoft and Amazon, started to sell space on their own, capacious server-networks, soon to be joined by hardware giants like IBM and Cisco, and large, specialist firms like Rackspace. Storage space is rented at so much per gigabyte per minute by users such as banks, health services, transport and logistics firms, companies with large sales forces; by advertising companies; by social media services (which make their money from the advertising companies); and from pornography sites which 'dwarf almost everything except the Googles and Facebooks of the internet'.[4]

Very early on, heat dissipation became a problem for data centers. Facebook's New York data center nearly caught fire during a period of peak demand in 2006.[5] Many data centers were moved to Iceland, with its abundance of glaciers and cheap electricity (Greenland has more glaciers and is closer to the US, but it is a much poorer country with only modest generating capacity). In 2012 Google took over an old paper mill in Finland, cooled by the sub-zero waters of the Gulf of Finland. At the same time, Facebook was spending $700 million on a data center in Luleå, Sweden, on a site just 62 miles south of the Arctic Circle.[6]

A 2012 study commissioned by the *New York Times* from the McKinsey accountancy firm found that most data centers leave their servers running at full stretch, even when demand is far below their capacity, just in case there is a surge in demand. Data centers were only using between 6 and 12 per cent of the power they consumed for data processing. They even kept diesel backup generators running at certain times of day, in case a surge in demand came when the public grid was also experiencing a surge. These have become significant sources of air pollution in California, and in upstate Washington, home of Microsoft's 'server farms'.[7]

In 2011 Britain had around 14,000 of these installations, with a peak consumption of around 6.44 GW ('enough to power six million homes', which is nearly a quarter of all the homes in Britain), and with an additional 25 per cent expansion the following year.[8] The *New York Times* study found that there were more than three million data centers worldwide, and that this was one of the few boom areas amid general economic slump.[9] The energy cost is even greater than these figures show, because it does not include the energy used by all the computers and other devices that access the cloud. A report mentioned in *Time* in 2013 says that as a whole (internet plus the computers that access it) the global ICT system used about 1,500 terawatt-hours per year, or about 10 per cent of the world's electricity.[10]

A substantial fraction of the energy used by people's computers is, again, the result of advertising. In 2010, two Dutch researchers measured the extra work a computer has to do, to download all those ads that appear in web-pages (typically, heavily graphical, and often containing animations). They did this by running the same, representative sample of web pages through two groups of computers for a whole weekend. An ad-blocking utility was installed on the first group of machines while the second accessed the pages normally and used 3.4 per cent more electricity. This suggests that Dutch internet users are using enough electricity to power 1,891 Dutch households for a whole year, or the output of 3.6 average wind turbines, just to upload advertising material they never asked for.[11] And this could be only a very small fraction of the extra energy use imposed on users by other people's promotional requirements: thanks to Web 2.0, almost every Facebook page, Amazon page or newspaper page that you access also draws in data from any number of other sources, continually updating it as you read – and all of this means that your computer does extra work.

These examples show how, once positional interests have been created, they combine in unforeseen ways, with impacts that are out of all proportion to the intentions of any single player.

FORCED MIGRATION: THE IRON LOGIC OF CORPORATE FLIGHT INTO THE CLOUD

The cloud's great attraction is that it is easier and cheaper for people and companies to store their data on a provider's internet servers than to store it on their own machines. Businesses can divest themselves of the headache and the cost of buying and maintaining expensive servers of their own (which take up office space); the headache of whether to upgrade them or not (if they do, the new ones might just sit around idle, while if they don't, they risk being caught out by a surge in business). The contract with the cloud provider will include a binding provision for risks like these – at a premium, which is however a known quantity, and so not a hostage to fortune.

Firms shed the burden of responsibility for their all-important data, the constant worry about obsolescence, of choosing hardware suppliers, employing and retaining staff who understand the 'dark arts' of setting up and managing servers, fixing them and so on; staff whose language they don't understand, whose word has to be taken to some extent on trust and who, if really good, might be poached by a wealthier rival.

All of these risks are avoided by 'migrating to the cloud'. Doing so may even be rewarded in the short term with reduced business insurance premiums. Failure to do so may be seen as a dangerous weakness by investors and their advisers, exposing the firm to the risk of a lowered share price and loss of credit-worthiness, which may help explain the staggering growth rate of cloud services during a world recession.[12]

The extraordinary energy needs of data centers (and therefore, income to be made from selling electricity to them) have dragged power utilities into the positional game. Even publicly owned utilities are now subject to 'market disciplines' and legally obliged to maximize profits and avoid financial losses, so they are in no position to refuse the kind of large-scale, round-the-clock custom the data centers offer. Hydroelectric, geothermal and nuclear plants make a particularly 'natural fit' because their electricity output can't be tailored easily to the ups and downs of normal domestic needs; a lot of it would otherwise go to waste during off-peak hours.

The *New York Times*/McKinsey report mentioned above[13] described how, in eastern Washington state, on the upper Columbia River, public hydroelectric utilities at first welcomed Microsoft's new server farms, but soon found themselves locked in a deadly embrace, having to dance to Microsoft's tune, and giving priority to the infrastructure Microsoft needed (like new access roads, sub-stations and power lines) over local needs.

One can also see how easily the needs of the cloud could become a new factor in the demand for new investment in nuclear capacity: nuclear plants do not cause the kind of obvious pollution that makes diesel generators so unpopular, and their relatively inflexible output is a less serious argument against them, if their surplus, off-peak output can be sold profitably to the data-center industry.

HOW THE WEB BECAME AN ENTROPY PUMP

An internet that uses this much energy might have been inconceivable had it not been for the positional pressures that fastened onto the Web almost as soon as it emerged from the labs at CERN (the European atomic research agency in Geneva) in 1991, leading to commercially oriented concessions by the mid-1990s, and then, in the early 2000s, the bundle of new techniques known as 'Web 2.0'.

The idea of using computers to help disseminate knowledge goes back at least to the 1930s and many possible methods have been considered and implemented to varying degrees, of which the Web is just one.[14]

In 1945, Vannevar Bush, an engineer and computer pioneer who became head of the US wartime scientific effort, wrote a widely read magazine article called 'As we may think', describing an imaginary device called the 'memex'. This would hold all your books and other references on microfilm or something similar; you could add your own notes using a stylus and, most important, draw connections between things. Bush envisaged 'a new profession of trail blazers, those who find delight in the task of establishing useful trails through the enormous mass of the common record'. His main concern was to support the work of drawing associations between ideas, so that they did not become trapped in intellectual backwaters like those of the discoverer of plant genetics, Gregor Mendel:

> lost to the world for a generation because his publication did not reach the few who were capable of grasping and extending it; and this sort of catastrophe is undoubtedly being repeated all about us, as truly significant attainments become lost in the mass of the inconsequential.

Bush's article is well worth reading, and is, of course, available online.[15]

The Web as conceived by Tim Berners-Lee at CERN was not the most visionary system possible for computerized publishing, and Berners-Lee did not see it as that. He was working in a fairly junior administrative role, with few resources, and simply wanted a quick and easy way to set up a system for scientists to share documents.

At that time, all sorts of adventurous ideas for 'hypertext' were being discussed in the computer community. The term had been invented in the 1960s by the famous independent computer pioneer Ted Nelson, and referred to texts that can be explored in different directions by following links of various kinds, rather than in a rigid linear sequence. Berners-Lee knew a fair amount about the various options, but he opted for a robust if slightly inelegant technique already well established in the print-publishing industries: SGML (see below). He did not intend his system to be the final word on hypertext. But as soon as his system became of interest to positional, market forces, that is what it became (which is why, as Ted Nelson is fond of saying, 'nothing endures like the temporary'[16]).

The first Web pages were very simple, containing text and no graphics. Pages contained links to other pages, and to files, such as graphic ones, that you might need to get hold of. These text-only documents loaded quickly, even on computer setups that would be considered completely archaic today. Then, as now, you needed a special piece of software called a 'browser' to read web pages. (In Berners-Lee's original setup, you used it to write them, and edit them as well.)

The system used what is called a 'markup language': a practice that derives ultimately from printing, where an editor with a blue pencil 'marked up' an author's typescript so that the typesetter would know which typefaces to use, which bits were headlines, and so on. It made a mess of the manuscript, but in the print industry this did not matter. As computers began to be adopted in the print industry, markup conventions were standardized so that typesetting machines could read them and format the text automatically. This was formalized as Standard Generalized Markup Language (SGML) in 1986 and Berners-Lee used that as the basis of his own system: HyperText Markup Language (HTML). 'Tags' like the ones used in SGML are inserted into the text, telling the browser how it should be displayed. The browser can recognize tags because they always begin and end with angle brackets. This is an example:

```
<h1>Shock, horror!</h1><p>The Web is
broken, <em>and I broke it!</em> Click <a
href="http://http://www.forbes.com/profile/
marc-andreessen/">here</a> for the full,
shocking story!</p>
```

Viewed through a browser, this becomes something like:

Shock, horror!

The Web is broken, *and I broke it!* Click <u>here</u> for the full, shocking story!

... depending on how your particular browser has been set up to display headings, emphases, links and so on.

One of many nice things about the markup system is that it can be read by different kinds of machines. For example, using suitably equipped computers, the text can be read aloud for blind and visually impaired readers. Extensions to the language even allow different voices to be used for different parts of the text, and to speak from different directions.[17]

But, as you can see, the tags also make the text unreadable *unless* you are viewing it through a browser – and this can be important. For example, if you want to refer to a particular document, there may be no 'clean', definitive version anywhere on the web that you can refer to; you may need to take a separate, cleaned-up copy of it (leading to endless proliferation of copies, in different versions). Speakers of other languages are forced to learn the particular subset of English that is used by HTML.

Ted Nelson advocates an approach whereby documents are sacrosanct. Any information or instructions (for display or interaction) are provided in an associated document, as are subsequent changes and comments. Nelson has called these associated documents 'layers' or 'filters'. One can envisage a filter-document that is simply a list of hyperlinks to be associated with particular bits of the underlying document (referenced perhaps by their co-ordinates: so many characters, or pixels, down; so many across). Any number of 'filter' documents may be associated with any other document, by anyone, and stored anywhere on the internet. People might even devise their own tools and conventions for writing filter documents (for example, people working in other languages and character sets). Documents might contain or even consist entirely of sections of other documents – Nelson calls this 'transclusion' – avoiding proliferation of copies. In Nelson's scenario, copyright would be maintained – and even paid for per word; he is a believer in intellectual property. Such a system would be an idealized version of the idea of 'canonical texts', familiar to literary studies (and this is reflected in Nelson's choice of the name Xanadu, from Coleridge's poem, to describe it).

Nelson and his widely scattered confederation of allies (in Japan, Australia, the UK and the US) have built working versions of this and other alternative hypertext systems, but they are limited by the fact that they have to be built on top of the now-dominant, hierarchical HTML system.[18]

HTML's shortcomings seemed a thoroughly acceptable compromise when it was first released but it started to become a costly one when the Web became an arena for competitive, commercial activity – and the sheer weight of investment meant there was now no going back.

This began with the release in 1994 of Netscape Navigator – a commercial spinoff from the team at the University of Illinois that had developed Mosaic, the first publicly available browser. Taken unawares by demand from the world of advertising (which had hitherto taken very little interest in computers), the original Web community of developers found itself outflanked by Netscape – whose founder, Marc Andreesen, succeeded in getting a free, personal-use copy of its browser distributed with the Microsoft Windows operating system, providing an instant, massive consumer base for web advertising and sales.

Navigator violated Berners-Lee's original conception by allowing web designers to include graphics in their pages via a new <img> (image) tag.[19] Next, it allowed a web page to create 'cookies' (small, temporary documents) on users' machines, to preserve information entered by the user between page loads; and the addition of a programming language to the browser, which designers could use to generate pages automatically, to perform calculations (notably, 'shopping cart' totals) and customize pages in response to users' input, pulling in information, such as the latest prices, from a database. The language was named JavaScript by Netscape's marketeers, to evoke associations with Sun Systems' powerful, general-purpose Java programming language, which was regarded as the most advanced language at the time.

All of these developments happened amid a general, unchallenged consensus that the internet was destined to have an entirely benign environmental effect. But the addition of graphics alone contained the makings of a serious Jevons Effect, as can be seen when we look at the makeup of a typical post-Netscape web page.

Marc Andreesen is now one of the world's richest men, so naturally the *Forbes* website has a page about him. This page is not very long, but the HTML page from which it's generated comes to 51 A4 pages of text (consisting of 8,506 words, or 88,928 characters). Most of this is automatically generated HTML markup, plus two or three Javascript scripts which control various visual effects, such as revealing and hiding graphic images when the user's mouse pointer passes over them.

Removing the markup and scripts reduces the document to 6 pages of text (1,220 words, or 8,128 characters). But most of this is promotional effort of one form or another: a few advertisements but most of it is verbiage intended to keep you within the Forbes website by drawing your attention to other exciting stories (to maximize advertising revenues, companies need to maximize the length of time users spend on their websites, and there are specialist companies who charge the advertising agencies good money for monitoring this). Eliminating the promotional text, one is left with less than a page about Andreesen himself: a 196-word profile (1,265 characters) albeit including a link to a fuller, separate page about him.

If one character of text requires one byte of computer memory, Andreesen's potted profile occupies just over 1k (a kilobyte) of memory, but its promotional accompaniments have multiplied that modest requirement 88 times. And this is just for the text.

The profile page also contains no fewer than 120 graphic images. These have all, no doubt, been skillfully optimized so that they take up the least amount of memory consistent with the quality Forbes readers expect. Even so, collectively they add up to 766.48k – three-quarters of a megabyte – and because of inefficiencies related to the minimum size of blocks of available memory and the limited amount of time even a fast, modern machine can spend optimizing memory allocation, my computer tells me that it has given them a megabyte (a round thousand kilobytes).

Well, one megabyte is not all that big by present standards, but the thousand-fold increase that it represents is consistent with the serious impact that the Web is now having, and even so doesn't tell the whole story. For example, most of the 120 graphic images are 'rollovers': when the user's mouse-pointer rolls over one of them, a larger version is summoned from a server somewhere and displayed to the user; a cascade of disk-operations and global network dialogues each with its own small, additional energy cost.

THE COST OF THE DOTCOM BUBBLE AND WEB 2.0

These new capabilities made the e-commerce phenomenon possible, and created waves of positional panic throughout industry.

Unforeseen consequences included the 'dotcom bubble' of 1997-2000, when speculative money poured into internet-based ventures. By the time the bubble burst, US internet businesses were collectively valued at $1.3 trillion or eight per cent of the entire US stock-market – but losses were much greater because so many internet-based companies were not stock-market listed, and because of the mortgages and ancillary business activities that had depended on the sector. The slump that followed took $5 trillion off US stocks.

Perhaps in response to the new, more precarious business environment, companies threw greater effort into 'building a web presence' than ever – this being seen as a much more controllable risk than investing in bricks-and-mortar businesses. Alongside this was the constant admonition from the business and technology media, from influential pundits, and even from government ministers and heads of state, that companies that failed to move with the times and establish a web presence were doomed. Many businesses (notably publishers and booksellers, in their desperate efforts to compete with the online bookseller Amazon) were forced to spend heavily on website development, at the cost of other aspects of their businesses. In other words, the shift to the internet was largely driven by positional concerns in Hirsch's strictly rational sense. It was a game they had to play, like it or not.

'Web 2.0' lubricated this shift. Essentially, Web 2.0 is the result of the bundle of techniques, initially implemented unilaterally by Microsoft and Netscape, but resolved finally by Google in 2004 for its Gmail service, and dubbed 'Ajax' the following year by a web-designer called Jesse James Garrett. Ajax is short for 'Asynchronous Javascript and XML'. (XML stands for 'Extensible Markup Language' and is a 'superset' of HTML, which permits adventurous designers to create their own tags for their own special purposes). Ajax allows a web page to be modified automatically while the reader is using it, without reloading the whole page – for example, 'breaking news', new chat windows, the constantly changing stream of advertisements and even mini-commercials at the side of the page.

Ajax made features like Google's 'live search' possible, as well as the whole phenomenon of 'social media', and the constant updates and general 'busyness' of major websites – as well as sophisticated web advertising. Think of the changing sequence of photographs at the top of a travel or news site; the advertisements seeking your attention, and very likely placed there according to some algorithm's assessment of your interests and needs – discreetly indicative of the torrents of data transfer that go on in the background, between servers dotted all over the planet.

Web 2.0 made online auction and booking sites practicable, and online banking. When you are buying things online, and change a single item in your 'shopping cart', the new total is shown almost immediately, without your having to wait a whole three seconds while the entire page is re-generated on the distant server, and re-loaded into your web browser.

We all recognize that 'a whole three seconds' is really not very long at all, and perhaps if we'd been consulted about it we wouldn't have demanded an overhaul of the internet's infrastructure on that account. We tend to think of these speed gains being self-evidently necessary, yet essential tasks (most obviously, typing) are done now no more quickly than they were when computers had hundredths or thousandths of their present speed. These tiny gains do, however, turn fairly humdrum tasks into highly engaging ones, and some writers are arguing that it may also make them addictive in a clinical sense,[20] which fits uncomfortably well with the often-voiced idea that a good computer application ought to be 'seriously addictive'. In the personal computer media, 'addictive' has long been the highest accolade a piece of software or a new device could earn, and describing something as 'seriously addictive' is much more likely to mean 'you really must stop whatever you're doing and buy this, now!' than 'don't touch this with a bargepole!'[21]

The question of why tiny incremental speed-increases might have a sinister effect is beginning to receive serious attention. It may have something to do with the seductive/addictive psychology of other activities that involve real-time interaction, especially when they require little physical effort. Addiction of any kind depends on swift neural feedback to whatever stimulus its user employs, plus 'pairing' of the stimulus with regular activities in one's daily life.[22] Cigarettes proved more addictive than other forms of tobacco because the smoke is inhaled, so that the nicotine reaches the brain almost instantly – more quickly even than when injected – and you can smoke as you work, walk, cook, drink, chat on the phone, and lie in the bath. This continual, intimate pairing of activity and reward created an attachment that became even more difficult to break than heroin addiction, according to various studies.[23]

Completing the more general cycle of effects and causes, Ajax set new and higher benchmarks for computer performance, which helped drive the demand for new and faster processors, faster-access memory and more of it, which in turn made it possible for games companies, for example, to use higher-resolution graphics, made video-on-demand more feasible so that it came to supersede video hire, and so on... To a large extent, the processing speed of modern computers is driven by the kinds of tasks modern computers are given to do, which are themselves set by the needs of manufacturers to compete with one another.

Web 2.0 was created under pressure from increasingly powerful e-business interests, against principled resistance from the World Wide Web Consortium (W3C), whose founding principles included a commitment to making the Web accessible to all, including the visually handicapped and the blind, and in as many different ways as possible. A great deal of patient, thoughtful work was being invested in web-design protocols that would ensure that pages could be re-rendered easily in different text-sizes and colors for people with limited vision, read aloud, and navigated by spoken commands, rendered as Braille, or be easily reconfigured for printing. Fragmented and constantly changing pages, and the advertisers' obsession with visual impact above all, threatened to make the vision of a universally accessible and reusable web impossible. Ajax was eventually accepted as a W3C standard in 2006, in recognition that it at least does not entirely destroy all hope of an information medium accessible by all.

As well as helping turn the 'weightless economy' into a major consumer of fossil fuels, Web 2.0 accelerated trends to shed staff, close branches, increase the value earned (and work done) for companies by their surviving employees – and helped to fuel hysteria over security, and the business interests that benefit from it.

For example, a huge surge of effort was thrown into 'virtualizing' what are called 'legacy databases': company databases of sales and stock data that had hitherto resided on mainframe computers, accessed by people working at office desks, often through dedicated terminals. By converting these databases into forms that could be viewed in a web browser, staff could be 'freed from their desks' and place orders and so on from clients' offices or from motorway service stations, hotel rooms or even from home. The advent of devices like the iPhone made this kind of thing at first 'sexy', then de rigueur. All manner of organizations moved online in this way. Governments put their databases onto the Web, as did infrastructure organizations, for example so that maintenance engineers could access the control panels of power stations and substations from home, or the road, in the event of brownouts.

All of which brought with it brand-new possibilities for subversion and crime, which in turn meant huge business opportunities for security firms – especially after the commencement of the so-called War on Terror in 2001.

In July 2013, a British government report on e-crime called it 'a bigger threat than nuclear war'[24], helping to make it what one PR source calls 'a potentially attractive investment opportunity for suppliers':

> North America [presumably the part of it containing the US and Canada] is set to spend US$93.6 billion on cyber security during the forecast period [2013-23]. Despite the scheduled budget cuts, Europe represents the second-largest market, with the total cyber security market valued at around US$24.7 billion.[25]

Many people are cynical about the need for these big expenditures, but not policymakers, as far as anyone can tell. Jeff Porten, a technology journalist, attended one of the large industry conferences on this subject in 2012, where:

> the category list of the exhibit floor reads like the signs at the Post-9/11 World Office Depot: Access Control, Biometrics, Blast Mitigation, Bullet Resistant Systems, Citywide CCTV, and so on. Browsing through the catalog, I found a full-page ad encouraging exhibitors to advertise in two security trade periodicals in India – 'a US$1 billion... huge opportunity'.

The event was addressed by the US Secretary of Homeland Security, Janet Napoletano, who justified the 'greatest threat we face' claim with the news that 'critical infrastructure attacks over the internet are up 17-fold'. As Porten pointed out, the actual number of these 'attacks' wasn't very great: 160 in 2011, and many or most of them might simply have been spammers' programs that endlessly do 'the equivalent of trying a door handle to see if it's unlocked [trying] millions of computers in numeric sequence, and happen to include "critical infrastructure" only by accident'.[26]

Porten thought the obvious solution was not to put such critical systems on the internet in the first place. The internet is an essentially open-access system. Using it for sensitive material had created a brand-new need for complicated security systems, which then presented interesting challenges to spammers, pranksters and crooks, so that yet more security would be needed. If remote access really is needed, why not set up a private network, or perhaps a specialist network like the secure, inter-bank ones used by ATM machines? But none of the security experts present raised that question.

1 Michael Lesk, 'How Much Information Is There In the World?' nin.tl/howmuchinformation Accessed 5 Feb 2014.
2 'All too much: monstrous amounts of data', *The Economist*, 27 Feb 2010.
3 Gary Cook, *How Clean Is Your Cloud?* Greenpeace International, April 2012. nin.tl/howcleancloud
4 Sebastian Anthony, 'Just How Big Are Porn Sites?' *ExtremeTech*, 4 April 2012, nin.tl/pornsitessize Pornography is thought to make up as much as 30% of the cloud. In April 2012, ExtremeTech, the online high-tech news-service, calculated that one particular site, YouPorn, accounted for almost 2% of the internet's total traffic, and 'There are dozens of porn sites on the scale of YouPorn, and hundreds that are the size of ExtremeTech or your favorite news site. It's probably not unrealistic to say that porn makes up 30% of the total data transferred across the internet. The internet really is for porn.'
5 James Glanz, 'Data Centers Waste Vast Amounts of Energy, Belying Industry Image', *New York Times*, 22 Sep 2012.
6 Jessica Twentyman, 'Datacentres: Risks and Wider Implications', *The Guardian*, 11 Jun 2012, nin.tl/datacentrerisks Accessed 1 April 2014.
7 Glanz, op cit.
8 *Datacenter Dynamics Industry Census 2011,* London, 2011, datacenterdynamics.com
9 Glanz, op cit.
10 Bryan Walsh, 'The Surprisingly Large Energy Footprint of the Digital Economy [UPDATE]', *Time*, 4 Aug 2013, nin.tl/Cloudenergy
11 RJG Simons & A Press, 'The Hidden Cost of Web Advertising', University of Twente, The Netherlands, 2010.
12 Data Center Dynamics, 'The 2013 Data Center Census Is Here!', datacenterdynamics.com | nin.tl/datacentercensus Accessed 18 Feb 2014.

13 Glanz, op cit.
14 For a good, if eccentric, explanation of all the main issues here, see Ted Nelson, *Geeks Bearing Gifts*, Mindful Press, 2008. Nelson, who coined the word 'hypertext' and has built many alternative hypertext systems, provides succinct and trenchant accounts of all the major developments in this area, and demystifies the computers. See also: Jakob Nielsen, 'History of Hypertext', nin.tl/hypertexthistory Accessed 17 Sep 2014; and Bob Hughes, *Dust or Magic*, Addison-Wesley, 2000.
15 Vannevar Bush, 'As We May Think', *The Atlantic*, July 1945, nin.tl/Atlantic1945
16 Nelson, op cit, p 99.
17 This is a feature of Cascading Style Sheets (CSS), explained at w3.org/Style/CSS
18 Ted Nelson's Xanadu and other projects can be found via xanadu.net
19 An 'image' tag might look something like this: <img src="http://www.myWebSite/pics/picOfMe.jpg" width="100" height="142" border="0">.
20 Tony Dokoupil, 'Is the Internet Making Us Crazy?' *Newsweek*, 9 July 2012, nin.tl/internetcrazy ; Bill Davidow, 'Exploiting the Neuroscience of Internet Addiction', *The Atlantic*, 18 July 2012, nin.tl/internetneuroscience
21 The adjective 'addictive' occurred 1,750 times on the MacWorld website, according to a simple Google search in September 2014, overwhelmingly as a term of high approbation. The noun 'addiction' occurred only 434 times, and two of those in the first page of results did refer to concerns about actual addiction, to the iPhone.
22 David Krogh, *Smoking: the artificial passion*, WH Freeman, 1991.
23 Anne-Noël Samaha et al, 'Why does the rapid delivery of drugs to the brain promote addiction?', *Trends in Pharmacological Sciences*, vol 26, issue 2, 82-87.
24 Charlie Osborne, 'Cybercrime "Bigger Threat than Nuclear War", UK Lawmakers Say', ZDNet, 30 July 2013, nin.tl/cybercrimethreat
25 'Global Cyber Security Market 2013-2023', PRWeb, nin.tl/cybersecuritymarket Accessed 18 Feb 2014.
26 Jeff Porten, 'Pondering Cybersecurity in the Real World', *TidBits*, 13 Sep 2012, tidbits.com/e/13257

11

'The least efficient machine humans have ever built': how capitalism drove the computer down a dead end

The typical modern computer is a rehashed version of a 1948 design that can only do one thing at a time. Most of the time, its vast internal resources do nothing except use electricity. Its premature dominance, and then sales pressure, drove the rich analog tradition – it was analog computers that put humans on the moon – to extinction and has stifled better architectures. But the ancient operating system in your laptop still contains a hidden trapdoor into the lost world of cheap, time-shared computing.

In the early days of the computer revolution, there was a much greater variety of types of computers than most of us now imagine or can conceive of.

For a start, as historian Martin Campbell-Kelly has recorded,[1] 'computers' were often people, working in large teams using elaborate paper-and-pencil systems, mechanical calculators and increasingly clever mathematical techniques. This kind of work could be very boring and certainly slow, but not without its own particular value: for example, in the social diffusion of ideas and skills throughout a whole group of people, instead of these ideas and skills being locked up in an inscrutable mechanical device.

Being a human computer doesn't seem very modern but, with more autonomy and less heteronomy, maybe the technological path would have produced more tools to support and enhance that kind of work. At any rate, it could be preferable, say, to a job in an Amazon warehouse, running around all day, under constant surveillance, with a pager and a barcode reader strapped to your waist, pulling boxes off shelves, against the clock, on a minimum-wage, zero-hours contract – which is a form of computerized work that a lot of quite highly qualified young people end up doing these days.

Right up until the 1960s, a good half of the computers in use in Britain and the US were very hands-on affairs – analog machines – but these have vanished almost without trace, replaced by digital machines, which work with discrete numbers (digits) rather than the continuously varying quantities that a typical analog machine uses.

The digits that a digital machine works with are the famous 'zeroes and ones' of the binary system, where the zeroes and ones are represented by differences in electrical charge (but almost anything else would do: many early machines used vibrations in columns of mercury to store numbers). Numbers and letters are stored as sequences of zeroes and ones, but the same zeroes and ones can also represent the logical statements 'true' and 'false' that computer programs use. It is therefore an extremely elegant system – but not the only one. There was also at least one 'ternary' machine, the Russian Setun (described later in this chapter). Charles Babbage's famous Difference Engine (see below) used decimal digits – not because he didn't know about other options, but because that's how the tables it was designed to calculate would eventually be printed. Eckert and Mauchley's famous ENIAC machine also used decimal numbers.

None of these different styles of computation is intrinsically better, or more advanced, than any other. The important thing is what's appropriate for the particular task in hand. When the options are reduced to a single technology, complications follow.

THE BURIED WORLD OF ANALOG COMPUTING

Analog computers can take a great variety of forms but their basic principle is that they provide a physical model of the system you are investigating or trying to control. Their results might be given as numbers (especially if connected to a digital computing device) but could be the position of a pointer, or a graph, or an adjustment of some other piece of equipment.

As recently as the early 1960s, around half of the world computer population consisted of analog machines and there was very little sense among computer professionals that this situation was ever likely to change. Yet, as James Small explains in *The Analog Alternative*, the entire tradition has somehow been almost erased from history within just a few decades.[2] It has been called technological history's 'great disappearing act'.[3]

Analog devices have an ancient heritage, which includes calculating devices such as the slide rule, which was invented shortly after the discovery of logarithms in the 17th century and was the workhorse of mathematical and engineering calculation work right up until electronic calculators arrived in the 1970s. There is a long tradition of navigational, gunnery and astronomical aids that includes the Persian and Arab astrolabe and a nearly lost family of complex, geared astronomical devices of which the first-century BCE Antikythera mechanism (dredged up from the Aegean Sea in 1901 but not understood until a hundred years later) is the sole, barely surviving example.

Analog devices often defy prevailing notions of what technology ought to look like. All and any materials can be grist to the analog mill. To take a recent (perhaps deliberately) challenging example, Adam Adamatzky and his colleagues at the University of the West of England's Centre for Unconventional Computing have built a variety of biological computers, one of which models the road networks of various regions and countries using slime mold (*Physarum polycephalum*), and offers useful insights into highway planning problems.[4]

The New Zealand economist Bill Phillips's 1949 'MONIAC' (Monetary National Income Automatic Computer) used water. Phillips had realized that the mathematical models John Maynard Keynes and Joan Robinson had used to describe economic systems were very similar to ones used in hydraulics (Phillips had previously worked on a large hydroelectric scheme). His computer modelled the flows of cash, credit, savings and investments in an economy with a system of plastic tubes, reservoirs, pumps and valves. Users could experiment with different levels of tax, investment, inflation, employment and demand, and watch the effects unfold (almost literally, 'ripple through' the economy). MONIACs were used in teaching and in research at the London School of Economics and several other institutions until Keynesian economics went out of fashion in the 1970s. The machine has tended to be mentioned only as a curiosity since then but, when demonstrated for the BBC in 2013, it seemed to model the lead-up to the 2007/8 economic crisis rather accurately.[5]

WHAT ANALOG COMPUTERS ARE, AND WHY CAPITALISM PREFERS DIGITAL ONES

Analog devices are defined in two ways: as physical models (or analogs) of the system you are trying to control or examine; or as devices that allow continuously variable input and output. This contrasts with digital devices, in which information is broken into discrete units below which no data of any kind is retained – as you find when you enlarge a digital photograph.

The traditional oven thermostat is a simple analog device that demonstrates both principles. It uses a bimetallic strip: one that's made in two layers from metals that expand by different amounts when heated. As the oven heats up, one layer expands more than the other so that the strip bends – providing a satisfactory representation of the oven's temperature. The gap between the oven's actual and desired temperature is a physical gap, which closes as the oven comes up to heat, illustrating one of the big advantages of the analog approach: the device can often be its own switch, and needs no additional circuitry or interpretive apparatus. It can also be made as accurate as it needs to be, and work almost perfectly at that level of accuracy. It is possible to design bimetallic thermostats that can detect the most minute fluctuations in temperature at the far extremes of heat and cold, but domestic cookers do not need this; they use relatively simple, highly reliable ones that use the optimal combination of metals (traditionally copper and iron) for detecting temperature changes in the range used in cooking (rather than the extreme ranges used in blast furnaces and cryogenics).

It also gives you an insight into the perverse way the very advantages of analog devices can undermine their appeal from a business point of view: a firm whose survival depends on profits may prefer to make a single device that can be used in all possible situations, with minimal tinkering, and mass-manufactured as a standard commodity (perhaps, also, under the protection of intellectual property law). There is also little glamor in a cheap device made from iron and copper so, were it to come to a sales pitch, the high-tech version made from exotic materials, which will work equally well in outer space or a nuclear reactor, might carry the day.

Analog devices also favor a rule-of-thumb or heuristic approach to problem solving that does not distinguish rigidly between theory and practice, designer and worker. Analog people tend to be both, which does not help them to fit in with a very managerial culture.

The thermostat also illustrates the analog tendency to produce designs that are optimal for their tasks and therefore cannot easily be improved with next-generation versions. As the hypertext pioneer Ted Nelson once said, 'good design is a pinnacle, from which it is downhill in every direction'.[6]

William Thompson's 1872 tide-predictor is an example of the intimate relationship that existed in the analog culture between advanced theory and practical engineering skill, and the resulting elegance it produced. Thompson (subsequently Lord Kelvin) developed a system of steel and brass wheels, pulleys, elegantly shaped cams, and polished steel and brass 'ball-and-disc integrators', to perform the mathematical transformations that the French mathematician Joseph Fourier had developed in the early 1800s in his own studies of heat transfer in metals. It could plot tides for a given location for a whole year in about four hours. Like so many analog devices, it is a beautiful piece of engineering (and can be seen working in the London Science Museum, not far from the MONIAC and portions of Charles Babbage's never-completed, but earlier, mechanical digital computer, the Difference Engine).

Thompson was limited only by the mechanical problem of physically linking integrators in series, and powering them, so that the output of one could be fed into another. This would have expanded the machine's capability from differential equations (which calculate rates of change) to the vast class of phenomena that can be described mathematically by second-order differential equations (rates of change in rates of change): meteorology, the growth and decline of living populations, chemical interactions, civil engineering (including the design of bridges and skyscrapers), physics and mechanical engineering, quantum mechanics, electrical engineering, and the late-20th-century sciences of chaos and complexity.

The US engineer Vannevar Bush (whose 'memex' idea was mentioned in the last chapter) used electrical power to solve Thompson's problem in 1931 with his 'Differential Analyzer' – the first computing machine to be described in the media as a 'mechanical brain'. This used electric motors and electro-mechanical relays to drive the integrators and had a massive iron sub-structure to eliminate vibration and reduce play between components.

The main impetus here was the need to optimize the design of electric power-distribution networks, and balance the constantly changing loads on them. Similar work was going on in Europe and the Soviet Union, where it fed into the broader work of economic planning, as I'll explain in Chapter 12. During the Second World War, military needs took over as the driving force: the famous codebreaking work; an escalating demand for artillery 'firing tables' (for the aiming mechanisms of a rapidly growing variety of different types of guns and shells, under all possible combinations of range, wind speed and direction); and finally the nuclear-weapons program.

In some of these areas digital machines were eventually able to deliver more quickly than analog ones, but in general, analog technology was much faster (instantaneous in many cases) and it began to evolve at a spectacular rate as electronic techniques became available. Thermionic valves (known as vacuum tubes in the US) replaced Bush's electromechanical relays. Valve-based operational amplifiers (op-amps) completely eliminated the physical power-transmission bottlenecks Thompson and then Bush had had to contend with, so that analog machines became an increasingly flexible, creative medium for engineers and scientists to engage in modelling and simulation.

James Small explains that, for many of the people who used them:

> The real appeal of electronic analog computers went beyond high computing speeds and the capacity for solving differential equations. It lay instead in their ability to incorporate traditional engineering design practices, characterized by graphical methods, scale models, empirical trial-and-error and parameter variation methods, in a new electronic form. By redefining scale-model building techniques and enabling dynamic systems to be simulated in real time, they enhanced traditional engineering design practices. Furthermore, electronic analog computation minimized the mathematical and emphasized the experimental and visual content of the design process. They provided a new design tool that helped engineers bridge the gap between the limits of theory and practice, and the complex real-world systems that they were constructing.[7]

Vannevar Bush was strongly of this persuasion. He maintained that a major reason for building his machine was so that 'one part at least of formal mathematics will become a live thing'. His biographer, Pascal Zachary, writes of Bush's satisfaction when:

> One operator of the analyzer, a mechanic untrained in mathematics 'got to the point where when some professor was using the machine and got stuck... he could discuss the problem with the user and very often find out what was wrong.' Bush believed that the 'fundamentals' of differential equations had gotten 'under [the mechanic's] skin'.[8]

Why did this thriving technology disappear, even from computer history? Most books on computer history written in the past two or three decades either do not mention analog technology, or mention it as a quaint and antiquated technology that was predestined to be eclipsed by digital machines, and which had more or less died out by the end of the Second World War. Yet, as Small and others have found, half of many firms' computing budgets went to analog technology in the 1960s and demand did not peak until the late 1960s. It was analog computers that put humans on the moon. An article from the electronics press from 1987, speculating on the strange decline of analog, says that there were then still three largeish firms making analog machines in the US, one of whose machines, the Electronic Associates Simstar, 'provides a 5:1 speed advantage, price-for-price, and a 10:1 cost advantage, speed-for-speed, over supercomputers'.[9]

Under heteronomy, the victory of one technology over another is very often a victory of propaganda, of sheer sales effort. Donald MacKenzie has researched and described various examples, including the saga that led to the replacement of electro-mechanical gyro-compasses with electronic laser ones after 1978 (when Boeing adopted them, for the 757 airliner).[10]

Throughout the 1960s and early 1970s, the electronics company Honeywell bet heavily on promoting the idea that lasers were 'obviously' better (because more scientific) than the electro-mechanical gyros that were already in use. The project only came to fruition thanks to close connections with government, the Cold War, and the decision to use the first laser-gyros in nuclear missiles (which had short flight-times, and large targets, so the huge 'drift rate' was not a problem). At that time, the electro-mechanical gyro industry was a mature one with its own large body of skills and supplier-networks, most of which would be dispersed and lost. The devices it made were small, accurate and highly reliable, and laser guidance did not achieve the same level of accuracy until the late 1980s – a decade after Boeing gave the idea its official blessing.

MacKenzie argues, through this and many other meticulously researched examples, that 'a technological trajectory is an institution. Like any institution, it is sustained not... through intrinsic superiority... but because of the interests that develop in its continuance and the belief that it will continue.'[11]

THE BLEEDING EDGE

CLOCKS: WHY TODAY'S COMPUTERS MOSTLY DO NOTHING, BUT VERY QUICKLY

It is assumed nowadays that any computing device, whether it's in a computer, a phone or some household device, is a binary digital one, with a 'system clock', a central processor (its 'brain'), and does its work one logical step at a time, at a rate laid down by the clock.

Clock speeds for the first personal computers in the early 1980s were typically 2 Megahertz (a work-rate of up to 2,000,000 binary steps per second); now they're at least a thousand times that speed, in the Gigahertz range, and this increase in speed is often presented as the main evidence of steady and amazing progress in computer development (often citing Ray Kurzweil's famous assertion, mentioned earlier, that if cars had improved at the same rate, they would cost a hundredth of a cent and go faster than light). But most of this speed increase is illusory.

Modern computers are limited (one might say crippled) by what was, in 1943, an elegant technique for managing the fragile hardware of the day: the idea of co-ordinating all the machine's components with a metronome-like timing pulse (the 'clock'). This may have been first invented by the British post-office engineer TH (Tommy) Flowers, while he and his colleagues were building the now-famous Colossus computer for breaking German high-command ciphers.[12] The technique was also adopted by John Mauchly and Presper Eckert for their much bigger ENIAC computer, completed in 1946, which played an important role in the design of the first hydrogen fusion bomb. The one-thing-at-a-time clock-driven approach is also a good way of keeping a conceptual grip on the complexity of problems that have to be broken down into enormous numbers of tiny steps.

'At the time, it was the right choice,' says a 2001 MIT journal article. 'The circumstances in which they had to design, using vacuum tubes and relay circuits, meant that they really couldn't build a reliable computer without a clock governing the whole thing'.[13]

The disadvantage of using a clock was recognized early on by John von Neumann and Julian Bigelow: no matter how much hardware you have in your computer, it can only do one thing at a time. Their machine (the IAS Machine or 'Johnniac') didn't use a clock, nor did many other early machines, which were 'asynchronous' or 'process-driven': as each stage of a calculation was completed, its result instigated execution of the next stage, using as much of the computer to do it as necessary. But the 'clocked' approach was what the famous ENIAC machine used (see Chapter 1) and made its way into the general-purpose computer architecture that John von Neumann sketched out in 1946 after meeting the ENIAC team, and circulated to anyone who was interested. The clock did the job, and the ENIAC's designers, Presper Eckert and John Mauchly, stuck with it when they set up the world's first computer company to sell their services to the US government. Investment then flowed into this architecture, and it can now be hard to imagine how things could be done any other way. But, as computer historian George Dyson says:

> The last thing either Bigelow or von Neumann would have expected was that long after vacuum tubes and cathode-ray tubes disappeared, digital computer architecture would persist largely unchanged from 1946... The global computer, for all its powers, is perhaps the least efficient machine that humans have ever built. There is a thin veneer of instructions, and then there is a dark, empty 99.9 per cent.[14]

The 'frozen accident' of modern computer architecture is one part of the answer to the riddle of why the 'weightless economy' turned out to have such a huge environmental impact. Computer scientists were well aware that better approaches existed and tried to develop them, but once market forces took an interest in computers, they seized the first one that offered competitive advantage and drove development down that particular avenue to the exclusion of all others. Investment followed, and turning back became impossible.

Investment poured, especially, into the globalized race to make smaller, faster chips to overcome the old architecture's limitations. In this race, firms could take full advantage of cheap fossil fuels, globalization and all the special concessions that could be wrung

THE BLEEDING EDGE

out of governments. This produced the 'Moore's Law' phenomenon described in Chapter 8. By the time alternatives had been readied for trial, almost always at government expense, Moore's Law had wiped out their design advantage. Maarten van Emden has written of how Japan's Fifth Generation Project, which could use large numbers of cheap processors, all working in parallel without a clock, 'was pronounced dead on arrival in 1992' when its first implementation found itself outpaced by conventional software, running on a new generation of Intel chips.[15]

The Inmos Transputer (funded mainly by the British government) met a similar fate. This was an extremely simple processor that could be used in large numbers to build 'massively parallel' computers, but apart from a few years of glory as the 'reality engine' of choice for Virtual Reality and 3D animation applications, in spacecraft, and for some sophisticated seismic imaging work, it never managed to penetrate everyday computing and ceased development in 1994.

The binary, clock-driven approach also became entrenched by developments in software, especially as this, too, became an area for commercial investment. Large communities of interest develop around particular languages, whether they are free and open source (like the original version of the C language) or a commercial language like Microsoft's C++ and its successors.

The very first computer languages were just libraries of code that could be loaded into the computer, laboriously, and then invoked with a kind of shorthand instruction set. Then came 'compilers', which allow the code to be written in a relatively human-readable form, and then automatically converted into a digital form suitable for the particular machine's architecture. Compilers increasingly came to assume a clock-driven architecture. Projects like the Transputer have therefore had the extra task of creating their own, parallel-processing versions of standard languages and the testing and training tools to go with them.

Then came 'structured programming'. With this kind of programming, the machine's memory is used to keep track of where all the constantly updating bits and pieces of data and program code are in the actual hardware. Each collection of data is stored under a single 'variable' name (for example 'foo', to use a programming textbook favorite) so that it can be summoned repeatedly (or 'iteratively', via an 'if-then-else' sequence of conditional commands, which can be read fairly easily by a human being). Algol was one of the first computer languages to do this, developed by an international team based in Zurich in the 1950s.

Before iteration, writing code to sort data (into numeric or alphabetic order, by category, by word frequency, by destination, and so on) was a mind-boggling task. The distinguished British programmer and computer scientist CAR (Charlie) Hoare had this job when, as a visiting student in Moscow in 1960, he worked on the problem of sorting word lists for machine translation from Russian to English. He invented the famous Quicksort algorithm to help him do this without the help of a structured language, and has described the great difficulty of explaining his discovery to his boss (at a small British computer firm, Elliott Brothers):

> *I said timidly that I thought I had invented a sorting method that would usually run faster than Shellsort [the previous standard technique, invented by Donald Shell], without taking much extra store [physical memory]. He bet me sixpence that I had not. Although my method was very difficult to explain, he finally agreed that I had won my bet.*[16]

He then stumbled upon the new Algol 60 language, which for the first time allowed him to see, fully understand, and communicate what he'd done. Hoare has described Algol 60 as 'a language so far ahead of its time that it was not only an improvement on its predecessors but also on nearly all its successors'.[17]

Quicksort is a 'divide and conquer' technique: it divides the task up into a number of little ones that can be tackled independently of each other. This means that it lends itself very well to use on parallel-processing computers (as envisaged in the 1980s, giving rise to projects like the ones mentioned above). Instead of a single central processor that handles everything, and a system clock controlling the inevitable queues of waiting data and instructions, you have lots of processors and no clocks, and calculations find their own way through the hardware, in their own time, to where the logic tells them to go. But the self-fulfilling prophecy of Moore's Law put paid to that, and Quicksort and its many variants now spend most of their time consuming processor-cycles, and dissipating a good share of the billions of watts of electricity that the internet consumes at any given moment.

Because it became 'the only game in town', the clock has become a growing annoyance for computer chip designers. However fast the chip is, it can only run as fast as its slowest component; and the clock is often the largest consumer of power on the chip. Chip designers try to work around this by building little 'islands of clocklessness' into their chips, but this is inelegant and the main problem continues to get worse. A special issue of the IEEE journal in 2005 said:

> The industry will soon be able to manufacture chips so complex that timing analysis and clock distribution will become completely intractable problems.[18]

SOVIET COMPUTING: DIVERSITY UNDER SCARCITY AND BUREAUCRACY

In the Soviet system there was little possibility of good ideas being ruined by runaway commercial opportunism; the challenge here was a different form of heteronomy, centralized bureaucracy, plus an utter dearth of resources. On the other hand, the situation was mitigated (and to some extent subverted) by a strong, informal culture of reciprocity[19] and, for better or worse, people were not under quite the same pressures to deliver results. Despite the utter devastation after the Second World War, Soviet engineers and mathematicians produced and sustained a fairly wide variety of computer types until the Central Committee decided in 1969 to abandon support for Russian and Ukrainian designs, and 'play safe' (as they thought) by buying IBM 360 mainframes. The Dutch computer pioneer Edsger Dijkstra had declared the 360 'a baroque monstrosity',[20] but the Central Committee was not to know that. In 1959, during the 'Khruschev thaw', a delegation of computer

'The least efficient machine humans have ever built'

scientists from the US visited computing centers in Moscow, Leningrad and Kiev, and were able to see five different families of machines, all of which had interesting strengths, and not all of which used clocks. One of these, the Setun, has gained some recognition in the West because it seems such a radical departure from our idea of what a digital computer is like.[21] Only about 50 of these machines were built but there was plenty of demand for them – word having spread throughout the Soviet Union mainly by word of mouth. But a plan to mass-manufacture Setuns in Czechoslovakia was scrapped by the authorities in 1965.

The Setun, developed at Moscow University and named after the tributary of the Moscow River that flows nearby, was a ternary machine: it used three electronic states to store data and perform logic instead of the two used in binary machines: positive charge, negative charge and no charge. Instead of bits, it had 'trits'. This is also known these days as 'null convention logic' and (as with clocklessness) chip designers find ways of smuggling it into modern chips to get things done faster – but there seems to be no possibility of an entire ternary machine.

Ternary architecture needs radically less hardware and allows various programming problems to be solved in a straightforward way.

According to NP Brousentsov, who led the Setun project in its latter days, all of this made for an impressively robust, cheap, versatile machine that was also very easy to manufacture and program. It needed only a tiny set of instructions (just 24) so it did not need an 'assembler' (the daunting, low-level programming/compiling system that allows hard-core programmers and engineers to control today's complex, binary hardware). As word got around, the team was bombarded with requests from all over the Soviet Union and many different kinds of institution, but they never managed to get official commitment to expand manufacture to meet demand, and Setun was officially killed off in 1965. Brousentsov felt that officialdom saw it as an 'ugly duckling'.[22]

The Setun story was almost lost until fairly recently – which gives some idea of the ease with which alternative courses of development can just vanish, so that the 'present blessed dispensation' seems like the only one that was ever possible. But this was only one of many routes that computing could have taken – and could be using, were it not for the overwhelming positional interests that are able to develop around 'winning' technologies when an economy becomes dominated by the business of picking winners. Professional interests, industrial capacity, education and training, supplier industries and component-manufacturing industries, all conspire to restrict the options, no less than Soviet committees did. As soon as a technology's original purpose becomes confounded with its role as a source of profit, all of these interests gather in a huddle around it, and their combined weight, as it were, deepens the 'well of attraction' they occupy so that everything within range slides into it.

The baleful 1969 decision to 'go IBM' was worthy of any corporate capitalist boardroom and led, as the Soviet computer pioneer Sergei Alexeyevich Lebedev warned that it would, to the worst of both worlds: inappropriate technology for the intended purpose, and no control over it either. By 1981 the USSR had become so dependent on Western technology (and on industrial espionage, to make up for its lack of local expertise and access to the support networks Western users enjoyed) that Western intelligence agents were able to sabotage important Soviet technical projects by feeding them false information.[23]

TIME-SHARING: ANOTHER ABANDONED ROAD

None of this is to say that clocks are fundamentally a bad thing; just that if it's possible to have too much of a good thing, capitalism will make sure you get it.

Clock-driven systems made a lot of sense when interactive, 'time-shared' systems appeared on the scene from the late 1960s onwards. Time-sharing was an idea that allowed a single computer (typically, a desk-sized minicomputer such as the Digital Equipment Corporation's PDP-11) to share itself among a number of users, by sharing out processing time to each in turn. By 1979 the latest PDP-11s had a clock speed of 3.3 Megahertz

(MHz) – or 3,300,000 cycles per second: about a thousandth of the speed of a modern machine but plenty fast enough to drive a plain-text display for each user without any noticeable delay.

Time-sharing was a very parsimonious arrangement: all that each additional user needed was a 'dumb terminal' (also known as a 'thin client') consisting of a plain text, monochrome display, with keyboard, and just enough electronics to handle communications with the 'host'. There was no needless duplication of equipment, and the bits of the system that needed frequent replacement or upgrading (in particular, the software) were all on the host, where they could be easily maintained and constantly improved. Fixes and new features became available immediately to every user, automatically.

Time-sharing became available over larger geographical areas through systems like Tymshare (in the US), that allowed a company's branch offices to use the HQ's mainframe, or one of Tymshare's own Tymnet machines. Companies could sell spare computer time on their own machines to anyone who needed it, wherever they were – using dedicated datalinks, or ordinary phone lines. This was the basis for a large amount of online activity before the internet became generally available in the early 1990s. France's famous Minitel system (launched in 1982 and only wound up in 2012) had millions of users, who could do most of the things internet users now do, with better security, and without having to buy, understand and continually replace their own computers: the terminal was supplied free by France Telecom instead of a phone directory. Nearly a million of them were still in use when the system was closed down. Most other countries had initiatives like these, with different characteristics and possibilities of their own.

During the 1970s, when personal computing was emerging among hobbyists and political activists, time-shared systems predominated in business and government settings, and it was here that 'hacker culture' evolved, often at night, when legitimate users were in bed.[24] Time-sharing ceased to be the dominant mode only after 1981 (when IBM launched its first Personal Computer, aimed at business users). From here on, it became clear that there was more money to be made by selling an entire computer to each user than by giving more people access to mainframe machines. The fact that each individual computer needed to be replaced as the technology changed and had to have its own complete complement of software, which would also need replacing, created huge new interests, with commensurate power to influence consumer demand and industry policy.

Time-sharing produced some of computerdom's most interesting innovations: computer games (like the original text-based game, *Adventure*, and the first graphic game, *Spacewar*) and a hugely influential system called *Augment*, designed by Douglas Engelbart to help people in different locations to work collaboratively in real time, via text, voice and graphics. Engelbart's project was subsequently taken over by Xerox's Palo Alto Research Center (PARC) and many of its ideas were incorporated into a system (the Alto) that then, famously, inspired Apple's Macintosh and Microsoft's Windows operating systems – which are in many ways caricatures of what Engelbart intended.

For example, while it's true that Engelbart's system introduced such now-standard items as the computer mouse, 'icons' and separate 'windows' for different documents, on-screen menus and so on, these were never intended to be all that a user might need to work productively with the system. The overriding aim was to support and develop users' skills in such a way that they could ignore the interface paraphernalia and focus on the mentally challenging bit: the abstract task itself, including all the complicated inter-dependencies of pieces of information that might be involved. Alan Kay (a leading figure in the development of personal computing, who worked at PARC) has described seeing users of Engelbart's system and its derivatives 'flying along through the stuff at several commands a second'.

The abortion that happened after PARC was the misunderstanding of the user interface that we did for children, which was the overlapping window interface which we made as naive as absolutely we possibly could to the point of not having any workflow ideas in it, and that was taken over uncritically out into the outside world.[25]

'The least efficient machine humans have ever built'

A great deal has been written about the inefficiencies of the kinds of computer interfaces that came to dominate as computers emerged into the consumer economy, but one of the most trenchant analyses was written in 2000 by the man who should have been proudest of Alan Kay's 'abortion': Jef Raskin,[26] who had led Apple Computer's Macintosh development team to great commercial success in 1984.

On the eve of the Mac's completion, and after many battles and compromises, Raskin walked off the project and immediately proceeded to design a vastly simpler system (called *Swyftware* in its first incarnation[27]) in which 'a small set of elementary actions and methods... apply to a broad and varied landscape of applications'.[28] It used minimal hardware and no mouse – just a 1977-vintage Apple II computer and a keyboard with two extra keys, 'leap' and 'creep'. It needed no special graphics, did not have separate applications (such as word processor, mail or data-handling) or even separate documents. Yet it allowed its user to do all these tasks, and many more – and to do so much more quickly than with the Macintosh. Raskin took his idea to the Japanese firm Canon, who released the finished 'information appliance', the Canon Cat, in 1987. Around 20,000 of these were sold but by this time a bandwagon had started to roll and the Cat, and *Swyftware*, became a footnote in computer history.

One of the Cat's selling points was the idea that the device itself was not that big a deal; the main thing was your data. There was even a slogan: your disk is your computer. In the colleges that adopted the Cat, students only had to buy their own disks, which held all their work. The students could put their disks into the nearest machine and all their work would appear instantly on screen, exactly as when they'd left off: a far cry from the arduous process that prevails now. But this was not an idea other manufacturers were likely to emulate, at a time when there was clearly much more money to be made selling people whole computers rather than mere floppy disks.

Inefficiency is a constantly growing presence as computing evolves within the competitive market. Cloud computing (the extraordinary impact of which was described in Chapter 11) is like a cockeyed version of time-sharing. But where time-sharing was parsimonious, the cloud could scarcely use resources more lavishly: every user needs a fully equipped computer, and a fairly highly specified one at that, plus high-bandwidth internet access, plus all their own software, which they have to keep up to date if they want to stay in the game.

As an additional irony, most modern personal computers, and even most mobile phones, even cheap ones, can now, in principle, do time-sharing (and are of course many times faster than the original DECs and Univacs of the 1970s). This is because more and more of them use an operating system based on Unix, which was originally created for time-shared machines. This is why, when you buy a new Windows, Apple Macintosh or Linux machine, you have to create your own 'Account' on it, in which to store your applications and files, and log into it at every touch and turn: the operating system still believes it's running on a time-shared machine, with lots of users, each with their own separate accounts.

If you know how to make use of this fact, you can connect cheap terminals and even old machines with good keyboards to your best computer and, in effect, have as many computers as you like, for the price of one. This possibility occurred to Jim McQuillan, a computer engineer in Troy, Michigan, in the late 1990s while he was setting up a system for a dentist's surgery. To his delight, it worked. He and a colleague, Ron Colcernian, then set up the Linux Terminal Server Project (LTSP) to promote the idea.[29]

Usage figures are not easy to get for a system of this kind, which simply distributes information rather than selling boxes, but a rough survey in the early 2000s found that LTSP then had around 1.5 million installations worldwide, mostly in schools. A little later, there were around a million in South Korea alone, but these figures have since been completely eclipsed by the take-up in Greece, where the financial collapse of 2008 and the austerity program have made parsimony and ingenuity essential for survival. Interestingly, the next biggest current user of LTSP is Nepal. South Africa, Spain and, of course, South American countries have also seen large-scale adoption.

'The least efficient machine humans have ever built'

By 2013 there were signs that even mainstream computer companies, like Dell, and some large users were taking a renewed interest in time-shared 'thin client' options.[30]

COMPETITIVE PRESSURE NARROWS ALL OPTIONS

Western computing pursued the 'clock-driven' path to the exclusion of other possibilities due to the competitive pressures created, first, by military demands (such as the race to build more and more powerful nuclear weapons and missiles) and then by commercial competition. By the late 1980s, a huge industry had built up around the self-perpetuating quest for greater and greater speed. With existing business and military markets reaching saturation (or at least, the fear that they might do so, or that growth might be derailed by a business downturn or an outbreak of peace and goodwill), effort flowed into supporting capital's age-old quest to eliminate humans from the value chain, and into the creation of consumer markets, especially in media, cannibalizing entire industries: audio, video, photography, telephony, publishing, printing, newspapers and TV.

The unpredicted emergence of cloud computing is just the latest result, and driver, of this feedback loop.

1 Martin Campbell-Kelly & William Aspray, *Computer: A History of the Information Machine*, Basic Books, 1996.
2 James S Small, *The Analog Alternative*, Routledge 2001, 2013.
3 CC Bissell, 'A great disappearing act: the electronic analog computer', paper presented at the IEEE Conference on the History of Electronics, 28-30 June 2004, Bletchley, UK, oro.open.ac.uk/5795/
4 A Adamatzky, *Bioevaluation of World Transport Networks*, World Scientific Publishing Company, 2012; see also the Center for Unconventional Computing website: uncomp.uwe.ac.uk/research.html
5 Mike Hally, *Electronic Brians*, Granta, 2005, pp185-206; Chris Bissell, 'The Moniac: A Hydromechanical Computer of the 1950s', *IEEE Control Systems Magazine*, Feb 2007; BBC Radio 4, *Popup Economics*, 6 Feb 2013.
6 Ted Nelson, *Dream Machines/Computer Lib*, joint edition, Tempus/Microsoft 1987 (*Computer Lib* originally self-published in 1974).
7 Small, op cit, p 4.
8 G Pascal Zachary, *Endless Frontier*, MIT Press, 1999, p 52.
9 'Whatever Happened to Analog Computers?' *IEEE Spectrum* 24, no 3, Mar 1987, 18–18,

10 DA MacKenzie, *Knowing Machines*, MIT Press, 1996.
11 Ibid, p 58
12 Gordon Welchman, *The Hut Six Story*, M&M Baldwin, 2000, p 179.
13 Claire Tristram, 'It's Time for Clockless Chips', MIT Technology Review, Oct 2001, nin.tl/clocklesschips
14 G Dyson, *Turing's Cathedral*, Knopf Doubleday, 2012, pp 274 and 301.
15 Maarten van Emden, 'The H-bomb and the computer, part II: the revolution that almost didn't happen' *A Programmer's Place* (blog), April 2013, Available at: nin.tl/VanEmden2013
16 Charles Antony Richard Hoare, 'The Emperor's Old Clothes', *Communications of the ACM* 24, no 2, Feb 1981, 75–83.
17 Charles Antony Richard Hoare, 'Hints on Programming Language Design', Stanford Artificial Intelligence Laboratory, Dec 1973.
18 David Geer, 'Is It Time for Clockless Chips?' IEEE Design and Test of Computers, 1 Mar 2005.
19 Alya Guseva, 'Friends and Foes: Informal Networks in the Soviet Union (1)', *East European Quarterly* 41, no 3, 22 Sep 2007, p 323.
20 Maarten van Emden, 'Dijkstra, Blaauw, and the origins of computer architecture', *A Programmer's Place*, 14 Jun 2014, nin.tl/vanEmden2014
21 Mike Hally, *Electronic Brains*, Joseph Henry Press, 2005.
22 NP Brousentsov, SP Maslov, J Ramil Alvarez, and EA Zhogolev, 'Development of Ternary Computers at Moscow State University', *Russian Virtual Computer Museum*, nin.tl/ternarycomputers Accessed 6 Feb 2013.
23 This was the subject of the 'Farewell Dossier' given to President Ronald Reagan by French President Miterrand in 1981.
24 There is a particularly rich literature on this period, for example: John Markoff, *What the Dormouse Said: How the Sixties Counterculture Shaped the Personal Computer Industry*, Viking, 2005; and a much earlier classic, Tracy Kidder, *The Soul of a New Machine*, Little, Brown, 1981.
25 '1995 Vannevar Bush Symposium: Closing Panel', nin.tl/VannevarBush Accessed 8 April 2014.
26 J Raskin, *The Humane Interface*, Addison-Wesley, 2000.
27 Swyftware ran on an Apple II computer; Raskin's idea was then taken up by Canon, who commissioned him to design the Canon CAT.
28 Raskin, op cit.
29 Linux Terminal Server Project, ltsp.org
30 Cliff Saran, 'PC Market Crashes', *Computer Weekly*, 11 April 2013, nin.tl/PCmarketcrash ; Steve Greenberg & Bobbie Jones, 'Why Should You Deploy Thin Client Devices?' *TechTarget*, nin.tl/thinclientdevices Accessed 18 April 2014.

12

Planning by whom and for what? The battle for control from the Soviet Union to Walmart

'Planning' has usually been taken to mean top-down, central planning within states and companies, and clear distinctions between those who make the plans and those who are planned for. This was almost as true for the 'free market' West as it was for the Soviet bloc. But both systems also relied far more than they cared to admit on bottom-up, mutual self-organization – a phenomenon recognized by socially engaged scientists during World War Two, and which flowered briefly post-War in the cybernetics movement. This failed to break the stranglehold of hierarchy, and the dead hand of top-down control remains – if in the shape of mega corporations rather than nation states. But the principles of cybernetics still have revolutionary potential.

In the 20th century, the USSR and the US projected themselves and were seen as the two opposing poles of human possibility, with 'Soviet technology' on the one hand, and 'capitalist technology' on the other. Yet concepts and tools that may prove key to securing our common future emerged through cross-border collaborations and cross-fertilizations that were often achieved in the teeth of official disapproval, despite big-power confrontation.

For better and for worse, the two superpowers also had more in common than they liked to admit. The authoritarian tradition of so-called 'scientific management' developed by Frederick Winslow Taylor in the late 19th century and implemented by Henry Ford, which seems so quintessentially capitalist, was thought to have radically progressive potential by many of Russia's revolutionaries. Lenin admired and promoted the examples of Taylor and Ford, and established an Institute for the Scientific Management of Labor in Petrograd.[1] The historian Richard Stites explained in his 1989 book *Revolutionary Dreams* that 'Taylorism' and 'Fordism' took deeper root, and earlier, in the Soviet Union than they did in many capitalist countries.[2]

In an article published in 1991 (but written just before the Soviet Union and the Communist bloc's abrupt collapse in summer 1990) the Nobel laureate Herbert Simon observed that a Martian, approaching the Earth, and viewing it 'through a telescope that reveals social structures' would be astonished to find how little of the world's work was being done through market transactions (visible through the telescope as red lines) and how much came under the domain of social organization (colored green). Simon's Martian had expected to see clear structural differences between the capitalist countries and the socialist ones. He could see none!

> No matter whether our visitor approached the United States or the Soviet Union, urban China or the European Community, the greater part of the space below it would be within the green areas, for almost all of the inhabitants would be employees, hence inside the firm boundaries. Organizations would be the dominant feature of the landscape. A message sent back home, describing the scene, would speak of 'large green areas interconnected by red lines.' It would not likely speak of 'a network of red lines connecting green spots'.[3]

Simon's question was: why do people work so hard? We're supposedly motivated by financial inducements but, in practice, this doesn't appear to be what's going on at all. Simon was drawing attention to the huge influence of social factors on productivity. Irrespective of the formal hierarchy within which it takes place, even a simple cash transaction involves 'a massive exchange of

Planning by whom and for what?

information in both negotiation and execution' within a richly textured milieu of shared knowledge, structures and trust.

Both the US and the USSR owed huge, unacknowledged debts to popular egalitarianism and solidarity, and these were sometimes the only things that kept the show on the road. In the Soviet Union, the seismic upheaval of Stalin's first Five-Year Plan, launched at the very same time as people suspected of having 'levelling' tendencies were being purged, could not have succeeded to the extent that it did without the grassroots communist initiatives that workers improvised among themselves in the form of production, wage-sharing and house-sharing communes. Richard Stites calculated that up to 134,000 such collectives existed during the most intense period of the Plan, but they had all vanished by 1932.[4]

In both societies, people's work was increasingly dictated and shaped by people other than themselves: the situation that the philosopher and economist Cornelius Castoriadis called 'heteronomy' (where decisions are made 'in the name of another' – as mentioned in Chapter 7). If Castoriadis had also had a Martian friend, equipped to detect autonomous and heteronomous power, it would have seen a situation almost identical to what Simon's Martian saw. In both political blocs people's lives were overwhelmingly being shaped by others, but with a little more wiggle-room in certain cases.

It is uncontroversial to deride the USSR's bureaucracy and cumbersome planning systems, but the difference between Soviet and capitalist management was to a great extent one of degree, rather than of essence. Heteronomous planning certainly did not vanish after the break-up of the Soviet bloc; it went from strength to strength, and computer design evolved along with it, making possible the kinds of global industrial, financial and regulatory enterprises described in earlier chapters, predicated on unprecedented inequality, and cranking it up to higher and yet higher levels.

The process has encouraged and been lubricated by a triple political dynamic: ever-fewer restrictions on international movements of goods, money and elites; ever-stronger protection for property rights (to more and more things, and extending across national borders); and ever-greater restrictions on the personal autonomy of workers and would-be workers – via border controls, surveillance and micro-management techniques in the workplace. These processes have been greatly assisted by post-1990 developments in computer techniques that were developed as much in the Soviet Union as in the West, generally with the best of intentions, but with too little appreciation of how they would play out in social landscapes distorted by unequal wealth and power.

In the 1960s and 1970s, there was great interest in Soviet ambitions to plan their entire economy by computer. In the late 1950s, the Soviet premier Nikita Khrushchev felt confident enough to predict that the USSR's productivity would soon 'bury' the West. But by 1980, critics were claiming that it would be impossible to plan an economy containing 12 million products because the calculations would take forever (this was the subject of the 'Soviet calculation debate', described below) but, in 2013, Toronto-based researcher Nick Dyer-Witheford noted that:

> by the mid-2000s Wal-Mart's data-centers were actively tracking over 680 million distinct products per week and over 20 million customer transactions per day, facilitated by a computer system second in capacity only to that of the Pentagon.[5]

Walmart has advantages the Soviet Union lacked: faster computers and a narrower remit. It is not attempting to supply all of its customers' needs, whatever it says in its advertising, and it can offload problems it would rather not deal with by subcontracting, outsourcing, offshoring, and so on (and there are computer tools that facilitate this – see below). But whether we are talking about Soviet planning or Enterprise Resource Planning (ERP), we are talking about planning from the top down, within a power hierarchy: heteronomous planning, where a special person, who knows all the options, plans for others, who have no or little knowledge or discretion in the matter.

THE BENEFITS AND DANGERS OF CENTRALIZED PLANNING

Soviet commitment to planning achieved some extraordinary things, but hopes of overtaking the West in economic terms came undone from the 1960s onwards. There are many factors to take into account here: the burden of maintaining extremely high military spending for one; for another, the very different conditions from which the US and the USSR embarked on their competition for world ideological leadership. The founder of the US Institute for Economic Democracy, JW Smith, has put it thus:

> If one must compare the former Soviets with the United States, one must visualize what the United States would have been like if, when its industrial development had just begun, powerful neighbors had destroyed almost everything east of the Mississippi and north of Tennessee, and 30 per cent of their prime labor force was killed in defeating the aggressors.[6]

... and Mikhail Gorbachev, thus:

> I recall my railway trips from southern Russia to Moscow to study in the late 1940s. I saw with my own eyes the ruined Stalingrad, Rostov, Kharkov, Orel, Kursk and Voronezh. And how many such ruined cities there were: Leningrad, Kiev, Minsk, Odessa, Sevastopol, Smolensk, Briansk [and] Novgorod... Everything lay in ruins... In the West they said then that Russia would not be able to rise even in 100 years.[7]

The USSR's reconstruction was a formidable achievement – yet it never used cash markets for more than a tiny percentage of economic activity. As Michael Kaser wrote in 1970: 'Soviet economics is the analysis of how the world's second biggest economy operates without a market.'[8]

Should the credit for these achievements go to Soviet planning, or to Soviet solidarity? From the triumph of the Bolsheviks in October 1917, the USSR's leadership was thoroughly committed to top-down central planning, but was sustained (as Richard Stites argued in *Revolutionary Dreams*[9]) by a powerful sense of solidarity and even of hope. Its successes in industry, health and war made it seem the obvious model for planning efforts worldwide. Both population and health recovered rapidly after the civil war and the Second World War.

All of this gave the USSR great moral power. Distribution of resources was socially planned, even if it was planned badly, and not left to the vagaries of the market that had created the horrors of the Depression in the 1930s. The countries that emerged victorious at the end of the Second World War had also proved for themselves that planning really did work. In Britain (as mentioned earlier) wartime planning had produced the greatest improvements in public health in the country's history.

A United Nations report of 1967 speaks of a general feeling that 'the national plan appears to have joined the national anthem and the national flag as a symbol of sovereignty and modernity'.[10] As countries in Africa and South Asia won freedom from the old European empires, many of them embraced the new techniques. I was once told by a veteran Indian computer scientist of groundbreaking work on queuing theory in which he participated in the early 1960s, using a computer with 2k of 'core' memory. The picture across most of the newly independent countries of Africa was one of rapidly rising living standards and productivity, thanks in large part to a commitment to central planning.[11]

Even the big market economies accepted that markets at least needed regulation. The distinguished Polish economist Oskar Lange (who spent much of his career in the US) spoke for a fairly large body of opinion in 1967 when he described the market as a 'cumbersome... old-fashioned' method of matching supply and demand, and even 'a computing device of the pre-electronic age'.[12] The OECD (known these days as the club of rich countries, and a bastion of market orthodoxy) had Cornelius Castoriadis, a Marxist, as its head of statistics until 1970.

Castoriadis had been a member of the Communist resistance to the Nazis in Greece during the Second World War but, anticipating a Stalinist takeover, fled to Paris in 1945. Acutely sensitive to the dangers, as well as the benefits, of centralized planning, he envisaged a future in which computers would be used to support autonomy – in 'plan factories', whose job was not to direct the economy, but to work out the implications of different scenarios, including the needs and preferences of members of the public. He described how such a system would work in a pamphlet in 1957, after Soviet intervention to crush a workers' movement in Hungary the previous year.[13] Castoriadis seems to have been unaware of it, but while he was drafting an updated version of the pamphlet in 1972, a system very similar in principle to the one he described was being built and used, in Chile (described in the next chapter).

Castoriadis hoped that bottom-up planning would lead to 'a real market for consumer goods', and a much greater diversity of choice than the 'all-or-nothing' situation prevalent under conventional market arrangements where, as soon as a product ceases to be profitable, it ceases to exist – a system (he wrote) that should be seen as 'an absurd way of settling this kind of problem anywhere but on the raft of the Medusa or in a besieged fortress'.[14] But apart from Castoriadis, few people in any of the advanced nations seem to have paid much attention to the possibilities of planning economies or business enterprises in any other way than top-down.

However, research by the historian Jonathan Coopersmith has shown that other options were considered early on in the USSR that might have set things on a different course. These were, however, abandoned with the steady rise of authoritarianism in the 1920s, and the decision to adopt the GOELRO plan for electrification of the Soviet Union.[15]

ELECTRIFICATION OF THE SOVIET UNION: HETERONOMOUS PLANNING BECOMES THE GLOBAL NORM

GOELRO is the Russian acronym for 'State Commission for the Electrification of Russia' (*Gosudarstvyennaya komissiya po elektrifikatsiiy Rossiiy*), established in 1920. It was also the trailblazer for the subsequent and characteristic system of 5- and

10-year plans, around which much of the international debate about state planning subsequently revolved. Lenin launched GOELRO with the slogan 'Communism is Soviet power plus the electrification of the whole country'. It inspired a story by Boris Pasternak, which in turn became the basis of Nigel Osborne's 1987 opera *The Electrification of the Soviet Union* – in which 'electrification' refers also to the ferment of optimism and excitement in which the project was carried through.

The Russian-born economist and Soviet scholar Alec Nove has written that GOELRO was originally conceived by an anti-Bolshevist, a brilliant engineer called Vasiliy Grinevetsky.[16] Grinevetsky was Rector of the elite Bauman engineering school in Moscow until just before his death from typhus in 1919; a Bauman Institute newsletter of 2005 tells us that it was he who developed the project-based method of engineering education, known as 'the Russian method', which became the model for engineering schools worldwide, including the Massachusetts Institute of Technology.[17]

GOELRO and its successors came to be seen as the quintessential expression of Soviet socialism: massive, heroic, top-down. It certainly showed what highly motivated people can achieve. But what was eventually achieved was not so much a radical departure from conventional engineering practice, as a heroic demonstration of what would thenceforth be the unquestioned global norm for national projects of all kinds, starting with power distribution: large, centralized generating plants, linked to industrial centers by huge networks of high-voltage supply lines. The US had two similar projects: 'Giant Power' in Pennsylvania and 'Superpower' in the northeastern states. These were not as large, but the principle not only remained unshaken, it became set in stone. Designing this kind of power network became a major focus for computer development between the two World Wars (the generation of electro-mechanical, analog computers known as 'differential analyzers' – mentioned in the last chapter).

However, GOELRO might have taken a very different shape. A rival approach, backed by the influential Bolshevik and art theorist Boris Kushner (who was for a while a member of the GOELRO

Planning by whom and for what?

committee), aimed to break down the sharp and ultimately fatal division between Russia's rural and urban economies by designing and supplying 'hundreds of small stations instead of waiting over a decade for regional stations because revolution in the countryside demanded action within months, not years. Only small stations could fulfil that demand'.[18]

Kushner's approach was what we might now call 'intermediate technology' or 'appropriate technology'. Standard designs for these smaller plants were to have been created, and manufactured either locally or at least within Russia itself. Jonathan Coopersmith has argued that, had this approach prevailed, it might have led the Revolution down a much less tragic path and avoided the brutal mass-collectivization of farms in the 1930s. But this approach appealed neither to the engineering elite, nor to the party leadership, especially after Stalin's rise to power in 1924.

The Oxford economist and Russian scholar Michael Kaser tells us that, even so, GOELRO produced a range of powerful ideas and tools, which Western governments and corporations eventually adopted themselves:

> *The year after the publication of GOELRO (with 10- to 15-year targets for the energy sector), its environment was changed by the introduction of a mixed economy in NEP [the short-lived 'New Economic Plan']. The electrification plan became a public-sector program for the development of an infrastructure under a market mechanism... [which] led the central planning organs to the discovery of control methods suitable for a mixed economy which were discovered in the West only after the Second World War. The appearance of the first empirical input-output table in the world [was in the USSR] in 1925; similar but updated tabulations underlay each of the plan 'control figures' published until 1929.*[19]

The 'control figures' mentioned by Kaser were intended for guidance, to give planners and plant managers a realistic basis for calculating needs and outputs. But their meaning changed in 1931, and they became hard-and-fast targets that managers had to meet by fair means or foul, or suffer the consequences – with the wasteful and scandalous consequences that became the stuff of Soviet folklore. Kaser notes that the phrase 'control figures' reappeared after Stalin's death for the 1959-65 Plan, marking a brief reversion from Stalinist 'directive planning' to the original 'indicative planning'.

Input-output analysis did not feature in Western thinking until 1941, when Wassily Leontief used it to help plan the wartime US economy. Leontief had left Leningrad in 1925 after falling foul of the increasingly dictatorial authorities. He eventually fetched up at Harvard and published his analysis of the US economy in 1941: *The Structure of American Economy, 1919–1929*. His department was kept busy applying the technique to supply problems during the Second World War, and then to planning for the economy's reversion to peacetime in 1945-6.

Since then, input-output analysis has been fundamental to every kind of economic plan. The politically conservative *On-line Library of Economics and Liberty* has a page on Leontief, which explains that:

> Input-output analysis shows the extensive process by which inputs in one industry produce outputs for consumption or for input into another industry. The matrix devised by Leontief is often used to show the effect of a change in production of a final good on the demand for inputs. Take, for example, a 10-per-cent increase in the production of shoes. With the input-output table, one can estimate how much additional leather, labor, machinery and other inputs will be required to increase shoe production.

Pencils, paper, and people with mechanical tabulators achieved wonders this way. Using new electronic computers, it was clearly going to be possible to correlate all the thousands of inputs of energy and materials for any given industry or enterprise with the outputs required of them, and those of other industries and firms. Surely this would be a far better way of doing things than trusting to markets, with their time-lags and wild oscillations between scarcity and glut.

LINEAR PROGRAMMING, WITH AND WITHOUT COMPUTERS

In the USSR in 1938, Leonid Kantorovich had discovered a mathematical method for getting maximum output from ramshackle collections of equipment – initially, machines for making plywood. The idea was to become known post-War as linear programming – another natural candidate for the new electronic machines, although Kantorovich showed that once you'd got the idea (and any factory manager could do this, he insisted) you could make a pretty good job of it with a slide rule.

Two others discovered the technique independently: a Dutch mathematician, Tjalling Koopmans, and a US mathematician, George Dantzig. Koopmans was working on the problem of assigning cargoes to ships, and the ships to convoys, with as little precious hold-space wasted as possible; George Dantzig's problem was how to schedule the training of thousands of air force pilots, and then how to plan optimal meals for them from whatever ingredients were available.

When the delegation of US computer scientists mentioned in the last chapter visited Leningrad in 1959, Kantorovich and his colleagues were applying linear programming to the problem of rebuilding rolling stock for the USSR's railways, achieving 'a great economy of metals', and on 'complex problems in the oil industry'. Most of this work was done on hand calculating machines, in factories where staff were being trained in linear programming. Fairly large input-output problems were being solved in this way (for processes involving combinations of 30 different inputs and outputs) and much bigger ones simplified. Digital computers were being used but, of necessity, human knowledge and skills had to be developed and explored first – evidently with greater success than one might have thought possible.

There's a suggestion, here, of two radically different notions of progress and two quite different paths that it might take: one where the new skills and concepts are diffused rapidly and widely throughout the population; and one where they are embodied in tiny metal and plastic packages, soldered into the bowels of an inscrutable device that only a tiny initiate can understand.

The visitors observed that, although the technical side of the Soviet effort was not as advanced as in the West, they were far ahead on theory, having 'more and better logicians and mathematicians... than at any universities in Western Europe or the United States', and 'Soviet use of computers may be expected to surpass in quality and quantity that of the United States, that is, unless the top-notch American mathematicians wake up to their responsibilities'. In their report, they quote a leading scientist, BF Semkov, to the effect that work already under way would soon make it possible 'to solve the questions of the largest rational co-operation between regions of the land, and to work out the optimal scheme of flow of goods'. [20]

The manifest power of this combination of technology and theory convinced many that the market's days were numbered. Indeed, right through the 1960s and 1970s, a very large part of the world's industrialized population grew from utter poverty and devastation to impressive strength without using the market at all.

THE CURIOUS INCIDENT OF THE CAPITALIST CALCULATION DEBATE

The Capitalist Calculation Debate is a debate that never happened – a significant non-event, like the dog that didn't bark in the Sherlock Holmes story, *Silver Blaze*.[21] The debate that *did* happen, and which rumbled on in socialist journals during the 1960s and 1970s, was the 'Soviet calculation debate', mentioned earlier. Was it possible to plan a fully modern economy using the mathematical techniques described above?

According to the economist Alec Nove, the answer was a decisive and scornful 'no'. His Ukrainian source (O Antonov) had estimated that planning the Soviet economy on input-output lines would involve calculating interdependencies between 12 million different products. Since any of these might be inputs to any other (Antonov calculated) it would take the entire population of the world 10 million years to produce a plan just for the Ukraine.[22]

In 1984, Cambridge economist Geoffrey Hodgson took a closer

Planning by whom and for what?

look and found that then-current computers could do the job much faster: a mere 18 years. But (he wrote later) 'what was overlooked was the reckless pace of development of computer technology'.

In 1993, two professors of computer science, Paul Cockshott and Allin Cottrell, pointed out a glaringly false assumption in the debate so far: not every product is needed in the production of every other product. No lipstick is involved in steel-making, for example. Therefore, the input-output table for any given product is rather small: what they call a 'sparse matrix'. This fact, plus the vastly lower price and greater speed of computers in 1993 already made it possible to manage a modern economy in something near to real time – and they describe in some detail a practical system of interconnected PCs, using ordinary spreadsheet programs, exchanging data on stock, output levels and so on every 20 minutes.[23]

When they were writing their book, Cockshott and Cottrell were unaware that a system using very similar principles, and very much simpler hardware, had been functioning in Chile during the last days of Salvador Allende's Socialist government, prior to the Pinochet putsch of September 1973 (described in the next chapter). A great irony is that linear programming and all the other mathematical tools envisaged during those early days, were set busily to work in the service of capitalism – and indeed proved essential for the creation of the modern, globalized, liberalized economy. Reincarnated in what are called Enterprise Resource Planning (ERP) systems, they are what make giant corporations like Walmart and Tesco possible.

Centrally planned economies now dominate the world economy, but instead of having 'Socialist Republic' after their names, they have 'Inc' and 'plc': they are massively centralized, commercial hierarchies, and this has happened without much discussion of their feasibility or otherwise.

Linear programming has gone from being a tool for superseding the wastefulness of markets into one that allows market players to operate on an unprecedented scale, competing against one another, and extracting value from suppliers and subcontractors more ruthlessly than was ever possible before.

Linear programming is used nowadays for such problems as maximizing the coverage of an advertising budget; maintaining the consistency of globally branded products from thousands of possible ingredients; and quickly finding the most profitable way of farming out the work of manufacturing running shoes or electronic goods between thousands of competing suppliers. 'Just-in-time' manufacturing and 'lean production', whose effects on workers are described so vividly by Paul Stewart and his colleagues in their book *We Sell Our Time No More*, would be impossible without it.[24]

It is even used in the elaborate 'combinatorial auction' systems that have been created for the auctioning-off of public assets – initially the sale of British public bus services to private companies in 1994, where the problem was to find the most lucrative combination of bids from many different purchasers for many different bus routes, each with different characteristics (some profitable and short, others unprofitable and long, and everything in between). The system has been hugely refined and elaborated since then, and was used to sell off the 3G electronic spectrum for mobile phones in 2000 (raising billions for governments but nearly bankrupting Dutch Telecom), and the 4G auction in 2013, in which each bidding company employed its own army of mathematical economists and game theorists, yet in the end raised only around a tenth of the 3G sale in Britain because of the recession. The auctioning of airline routes uses the same kind of technique. The companies that have developed these systems have done well, exporting them to countries that are starting to privatize their own public services (the London-based management consultancy Coleago is one of these).[25]

CONNECTION-MAKING AND THE ECOLOGY MOVEMENT

The planning techniques described above contributed to a much bigger movement that was taking shape in bits and pieces and in different disciplines all over the world during the 20th century.

It recognized two fundamental truths: first, that everything is connected and, second, that everything has its own perspective, its own experience and intelligence, which cannot be ignored. Arising from these is the concept of 'emergence' – the recognition that complex and robust behavior can develop from apparently simple organisms without outside intervention, and that this behavior is likely to be much more efficient and robust than anything specified and controlled from outside (by 'higher authority').

The connection-making tendency can be seen very strongly in the ecology movement's emphasis on the interconnectedness of life and the need to treat the biosphere as a single system. Input-output techniques have become an essential tool here, and are central to the kind of analysis Eric Williams did (see above, Chapter 9) of the full bill of materials for making microchips.

It also surfaces in the new disciplines that came to the fore during and just after the Second World War: Operational Research (OR), General Systems Theory and Cybernetics[26] (as well as some less well-known movements like the British Sociotechnical Systems movement, and the Participatory Design movement, which was particularly dynamic in Scandinavia). All of these movements involved crossing disciplinary boundaries that often seemed sacrosanct to those in authority, and a general preference for multidisciplinary working: psychologists, chemists and biologists were as likely to be part of the team as engineers or mathematicians.

It affected thinking about education (a growing emphasis on multidisciplinary curricula), health (holistic and community-health approaches, and epidemiology), and the study of history where, starting with Marc Bloch (whose 1935 work on medieval water wheels was mentioned in Chapter 2), there was a growing insistence on treating history as a whole, freed from the national story-telling tradition that had dominated the discipline previously. Bloch wrote: 'the only true history, which can advance only through mutual aid, is universal history'.[27]

Bloch and his successors (Fernand Braudel in post-War France, Immanuel Wallerstein in the US) began to expose the illusion that history is shaped by some grand, evolutionary plan. History is always contested at the point where it is being made, which is to say, everyday life. Bloch practised what he preached, ending his career as a somewhat elderly resistance leader, killed by the Gestapo in June 1944.[28]

The importance of 'minor players' was a particular study of biologists and physiologists such as the Estonian/German physiologist Jakob von Uexküll, who, in the 1920s, studied how such apparently simple organisms as jellyfish, amoebae and insects perceive their environments (their 'umwelts' – perceived worlds); how they get and deal with feedback from it, and the surprisingly complex and adaptive behaviors that emerge.[29] Von Uexküll's work demonstrated the importance of taking a 'bottom-up' view of the world and was influential in various other fields, including linguistics – as well as in the more politically sensitive matters of work organization.

All of these various methods of enquiry insisted, while respecting the 'atoms' of the situation, on dealing with the *whole* situation: not just the animal and its food, or a machine and its operator, but the total environment insofar as it affects the object of interest. This would lead to radical conflict in the world of industry or politics, where it comes face to face with managerial desire to monopolize the 'big picture'. But that monopoly broke down somewhat during the Second World War, when the forces protecting old inequalities had to take second place to collective effort. This allowed the new approaches to break out of academia and into public life.

OPERATIONAL RESEARCH

Operational Research (OR) emerged in Britain in the 1940s from the urgent need to tackle real threats, from the daily sinking of ships to the prospect of starvation. One of OR's pioneers, Stafford Beer, wrote later that it's untrue to say that nothing good ever comes from war: 'there were then whole areas of military management in which circumstances compelled those in command to acknowledge freely that they were guessing'.[30]

OR accommodated all kinds of people. Churchill, the passionate anti-communist, found himself tolerating and working with a fair number of communists and socialists. Leading government advisers like Solly Zuckerman and Patrick Blackett and many of those who worked with them were left-liberals or socialists; JD Bernal (who masterminded the successful D-Day landings in France in June 1944) was a lifelong communist.

As scientific adviser to the Admiralty, Blackett developed the careful data-gathering approach to major problems characteristic of OR, starting with the loss of shipping from U-boat attacks. 200,000 tons of shipping were being sunk every month by April 1941, and the amount would rise to 600,000 tons a month by the summer of 1942. Blackett's methods, involving large numbers of structured interviews with aircrews, survivors, engineers, plus first-hand observation of the problem from an aircrew perspective, resulted in a 700-per-cent increase in the number of U-boats sunk. Stafford Beer attributes this to a simple adjustment to the settings of depth-charge fuses[31] – but new radar techniques were also involved, which put the Admiralty in competition with the Air Ministry for this scarce new resource. Survival rates of British ships were also helped by Blackett's discovery that larger convoys were safer than smaller ones.

As Stafford Beer pointed out later on, decisions can never be based simply on facts. In the event, action has to be guided by beliefs. Facts are never gathered for their own sake, no matter how coolly objective one may claim to be, but always and inevitably to confirm or preferably to challenge pre-existing hypotheses and beliefs. One can never know everything. The best anyone can hope for is that one's beliefs are reasonably well founded.[32]

When it came to advising on British area bombing policy, the OR people were up against beliefs that could not be shifted by factual evidence. Using the same methodology as with the U-boats, Bernal and Zuckermann, with 40-strong teams of interviewers, surveyed all aspects of the effects of German bombing on Birmingham and Hull and found that it had had surprisingly little effect – either on production or on morale. More workers turned up for work during the raids than normally.[33] However, Churchill and his personal adviser, Frederick Lindemann, could not be shifted from their belief that heavier bombing would break German civilian morale. Blackett responded that a better idea would be to break the German High Command's morale by destroying their supply depots and war materials. Of the bomber war, he wrote: 'So far as I know it was the first time that a modern nation had deliberately planned a major military campaign against the enemy's civilian population, rather than against his armed forces.' He also calculated that the diversion of resources into the bomber campaign had lengthened the war by at least a year.[34]

Blackett blamed himself for failing to carry the argument, but recognized that the milieu in which it had taken place was more to blame: the world of entrenched inequalities evoked by the phrase 'corridors of power'. Much bigger reputations and budgets were at stake than had been the case with the U-boat emergency – even though the latter had been far more relevant to the country's survival. Blackett would go on to become a strong opponent of nuclear weapons and an advocate of development aid for ex-colonial countries.

As the work of OR developed, its practitioners came more and more to see its relevance to society in general. RW Bevan, the British Air Ministry's Senior Scientific Adviser, and one of the pioneers of wartime OR, wrote in 1958 that:

> ...operational research should be applied essentially to broad problems, as for example to an industry as opposed to the bench, to a transport system as opposed to a railway truck... Incidentally, I can see no reason why operational research should not be applied in the future to studies concerned with national planning and national policy.[35]

Stafford Beer would have an unexpected opportunity to do this in 1970 (see next chapter) but meanwhile the ethics and methods of OR, combining with the cybernetics movement as it developed in the US around the mathematician Norbert Wiener, infected all manner of areas that had never been opened to this kind of analysis before.

The related 'sociotechnical systems' movement, which emerged during the War from the Tavistock Psychiatric Institute, examined the ways workers tackle the challenges of their work, and the vital importance of self-management in teams. Eric Trist is credited with creating the field – initially to examine military teamwork, but he soon realized its much greater scope in industry (most of which was by then in public hands). Enid Mumford, who later became a leading figure in the Human-Computer Interface (HCI) community, wrote a brief history of the movement in the 1980s, in which she describes a study she was engaged in (while at Liverpool University in the early 1960s) into the psychological impact of a new coal-getting technology (the 'long wall' method), and the unexpected ways in which workers organized themselves to take control of the new system.[36]

Like many in 1980, Trist saw the potential for revolutionary change in the whole structure of society:

> Advanced Western societies are on the threshold of a profound change in the texture of their socio-technical relations. This entails a change not only in quantity but in quality. It represents a discontinuity, as witness the opportunities for scaling down rather than up, dispersal rather than concentration, and self-management rather than external control. For the first time since the industrial revolution, a major class of technological forces is supportive (in potentia) of efforts to countervail some of [its] main negative impacts.[37]

But in the post-War years, old hierarchies reasserted themselves. OR practitioners found less and less scope for tackling 'whole problems' and more and more pressure to come up with fixes to local problems, ignoring their context. Stafford Beer was complaining in 1966 that while OR had been done for individual departments in the NHS, transport and power, it was not being done for the entire health, transport or power system – let alone for the entire economy. Even the military (early if reluctant adopters of OR during the War) insisted on being three separate, rival systems (army, navy and air force). 'Once the powerful ecological threat of extinction is removed,' he wrote, 'civilization collapses into its sub-systemic attitudes'.[38]

OR was strong on both sides of the Atlantic and for a time seemed to sing from the same hymn sheets but, as Paul Edwards has shown, by the late 1940s it had started to morph into something less holistic and more business-friendly, under the umbrella term 'Systems Analysis', at the RAND Institute. Whereas OR sought to address problems by optimizing the use of existing equipment, Systems Analysis saw its role as specifying new equipment to deal with problems. It coined the term 'cost-benefit analysis'.[39]

US Operations Research nonetheless had its leftwing moments. In 1968, the Operations Research department at the University of Pennsylvania in Philadelphia made its facilities and staff available to the local black, ghetto community of Mantua, to use as they saw fit. As the project's instigator, Russell Ackoff, records, this grassroots project grew by 1971 into a $6-million-a-year operation employing just over 300 people, repairing local properties, planning new housing, with a community service-station for bikes, cars and domestic appliances, two newspapers and a local radio station.[40] The project's motto was 'Plan, or be planned for'. A little earlier, in 1966, the Black Panther Party concluded its list of 10 demands with 'people's community control of new technology'.[41]

In general, however, it seems that the US school of OR was more readily incorporated into US business and governmental culture, possibly because of the greater sophistication of US management (Frederic Winslow Taylor's gospel of 'scientific management' had made great headway in the US between the World Wars and relatively little in Europe). Robert McNamara, who became US Defense Secretary and was responsible for the destructive yet ineffective military policy in the US-Vietnam war, had found his route to power through OR at the Ford motor company – and used arguments from Operational Research and cybernetics to justify the US bombing policy.[42]

CYBERNETICS

Cybernetics emerged separately in the US in the last years of the Second World War, around the mathematician Norbert Wiener, and came to embrace all of the other traditions mentioned above to a greater or lesser extent. Wiener formulated his approach as a result of and in reaction to his war work (automatic aiming systems for anti-aircraft guns) and his horror at where he saw the world heading.

It was Wiener who coined the word 'cybernetics' from the ancient Greek word for a helmsman: *kybernetes*. He published the founding text of the cybernetics movement in 1948: *Cybernetics: Decision and Control in the Animal and the Machine*. He followed it in 1950 with *The Human Use of Human Beings: Cybernetics and Society*, in which he warned that the greatest danger now facing humanity was a capitalism equipped with the new computing technologies.

The central interest of cybernetics was the way living systems of all kinds – individual organisms, and communities of organisms – maintain 'homeostasis', a reasonably stable 'internal milieu' even amid wildly fluctuating, hostile environments.

All of these systems use feedback from their environments to maintain homeostasis. Negative feedback can prevent a system going out of kilter by 'damping' whatever evoked the feedback (in the way the centrifugal 'governor' of a steam engine closes the throttle automatically if it spins too fast). Positive feedback, on the other hand, amplifies the tendency that evoked it, as when a small movement of surprise in a flock or a herd gets picked up by the other animals, and turns into a stampede, or into a stock-market 'bubble'.

The inequality-driven, positional phenomena described in this book are all examples of positive feedback, from the self-reinforcing tendencies that gave rise to the automobile economy and suburbanization, to the runaway development of computers along the trajectory that gave us Moore's Law and the throwaway iPad.

Cybernetics was both multi-disciplinary and international

from the word go. Wiener's first collaborator was a Mexican heart surgeon, Arturo Rosenbleuth, and the movement rapidly captured the interest of electronics engineers, anthropologists (like Margaret Mead and Gregory Bateson), psychologists, psychiatrists and biologists.

Cybernetics (like everything else) eventually became incorporated into military and commercial activities of which Wiener would have despaired and, in dumbed-down form, even became part of the promotional jargon of high-tech capitalism (as Richard Barbrook describes in his study of the 'culture wars' that ran in parallel to the Cold War[43]) but nothing changes the fact of its revolutionary potential for organizing and maintaining the stability of a more just, lower-impact society.

This potential is explained in a handful of concepts developed by two of Wiener's friends, the psychologist W Ross Ashby and the management consultant Stafford Beer. These concepts have been found to be as applicable to societies as they are to machines.

ULTRASTABILITY AND 'REQUISITE VARIETY'

Before he became directly involved in cybernetics, Ashby amused himself by exploring the concept of homeostasis, which he did by building a 'homeostat': an electromechanical device, built from RAF bomb components, that could tolerate being shoved around, quickly restoring its own internal electronic state to within certain parameters. Small communities of these devices could be unleashed on each other like fairground dodgems, without the situation degenerating unduly.[44] One of Ashby's many insights was:

> There can't be a proper theory of the brain until there is a proper theory of the environment as well... the subject has been hampered by our not paying sufficiently serious attention to the environmental half of the process... the 'psychology' of the environment will have to be given almost as much thought as the psychology of the nerve network itself.[45]

Planning by whom and for what?

Ashby identified the quality known as 'ultrastability'. A system is ultrastable when it 'can survive arbitrary and unforeseen interference',[46] and the world contains many examples of this. Spinning tops and gyroscopes may wobble if tapped, but then return to the same, steady condition as before. It is almost impossible to straighten a river that is determined to meander: however much earth-moving equipment you employ, it always reverts to a meandering course. Unjust societies can become frighteningly ultrastable, their leaders not only surviving attempts to overthrow them, but becoming more firmly entrenched, so that it can take a total catastrophe to change things.

However, ultrastable systems can sometimes be persuaded to 'jump' into a new and different ultrastable state. When a speedboat's hull lifts at a certain speed, it can stay in that attitude, despite hefty collisions with waves; but when its speed drops, so does the hull, which then becomes stable in the way an ordinary boat is, and rides the waves in a way more suited to low-speed maneuvering. Another way of describing this change of state is to say that the system enters a different 'phase space'.

Once in a particular state, a society's internal dynamics will tend to keep it there. The challenge is how to facilitate a social change from an unsustainable social order to a sustainable, egalitarian one. An encouraging yet tantalizing thought is that, in nature, these transitions from one phase-space to another are quite rapid.

Ashby also discovered and explained the Law of Requisite Variety, which says that a control device should have at least as many possible states as the system it is meant to control. Ideally it should have more.

If one thinks of everyday life in these terms we can see how woefully inadequate most people's control options are – and how easily their desires are overridden by a few people who have a superabundance of control over their own and other people's lives. For the majority, any hope of controlling the politics that shapes their lives boils down to voting in the occasional election or, if sufficiently provoked, going out on the street and making a noise, and possibly getting arrested or injured. For wealthy people, on the other hand, the main control device is their own social milieu, a capacious affair with an enormous repertoire of tools at its disposal: invitations to weddings, directorships, memberships of advisory boards of charities and the like, mutual favors. Very little effort or risk is entailed: this apparatus even responds to small acts of kindness, frowns and discreet throat clearances, quiet words, firm handshakes and compliments.

The Law of Requisite Variety has lots of applications. The radical architect John Charlewood Turner applied it to housing.[47] Within modern countries people have less control over how they are housed than one might imagine, especially if they 'play by the rules' and accommodate themselves in the approved manner, by choosing from the tiny range of legal options, instead of squatting or putting up shacks. House, he said, is properly a verb: 'the process of putting a roof over someone's head' but it has been turned into a noun, a thing, a commodity: a standardized item that you can buy more or less of at a given price.

A housing system that had requisite variety would give people as much control as possible of the resources needed for putting roofs over their heads, including access to the necessary skills, equipment, supply chains and systems for infrastructure planning. This would include support for negotiating the different needs and desires of neighbors and other local interests – complex combinations of potentially conflicting needs, no doubt, but people always somehow managed this in the past, and thereby created exactly the kinds of environments where (as Turner wryly observed) architects and the owners of big building firms prefer to live, such as pretty, traditionally built country cottages and Georgian terraces. These are also exactly the kinds of constraint-balancing problems that computers were designed to deal with (as Cornelius Castoriadis recognized).

Ashby's law even offers a physical explanation for people's capacity to stay sane and functioning when they are so very much at the mercy of forces they do not control. As another distinguished early cyberneticist, Heinz von Foerster, pointed out, our internal milieu (our conscious mind, plus all the unconscious processing the central nervous system does) is about 100,000 times richer than our sense-experience of the external world:

'Since there are only 100 million sensory receptors, and about 10,000 billion synapses in our nervous system, we are 100,000 times more receptive to changes in our internal than in our external environment.'[48] This rich internal milieu goes some way to explaining our ability to 'make a heaven of hell, a hell of heaven'.

VARIETY ENGINEERING: THE DIFFERENCE BETWEEN AMPLIFICATION AND SHOUTING

Sometimes, the amount of variety in the system is just too much to cope with and must be reduced. Engineers call this attenuation, or damping. The art of attenuation is to reduce the range of stimuli the controller or person has to deal with, while preserving just the most significant ones, which requires a richly varied response (as Ashby said: only variety can control variety). Air traffic control systems are full of that kind of circuitry, so that controllers do not get distracted from significant developments by all the activity shown on their screen. Conversely, significant tendencies that might not show up amid all the noise need to be identified and then amplified: some medical scanners and imaging systems do this.

Of course, what counts as 'significant' is defined by whoever specifies the system. There is little disagreement in the matter of air traffic control or brain scans, but enormous differences when it comes to managing economies, communities and firms, when power is held unequally. However, when power isn't held unequally, or there is a consensus that it shouldn't be, then impressive things can be achieved relatively easily in the general economy as well – as with Chile's Cybersyn project, described in the next chapter, which demonstrated that very simple computer hardware could help achieve radically improved efficiencies in meeting a society's needs, if it is serious about addressing them.

Under 'business as usual', however, managements are less concerned with discharging an organization's avowed purpose (providing nutritious food, easy transport between A and B, warmth, comfort, and so on) than with discharging their responsibilities to shareholders (to provide healthy dividends and an ever-rising share price) or to themselves (to maintain their careers on a constantly rising trajectory and their children at private schools). Such organizations are simply not equipped to tackle their alleged aims, but Ashby's Law decrees that the discrepancy between claim and reality must be made up somehow. Stafford Beer (chief architect of the Cybersyn project) once wrote that the conventional way to do this is 'essentially by shouting':

> This is not to confuse sound amplification with variety amplification. By declaring a monumental platitude or a crude oversimplification with conviction and an air of authority the politician seeks to convey the impression that since obviously nothing could be that naïve, it isn't. There are hidden depths in the underlying conceptual model, comes the implication, which you would not understand, or which it is not in the national interest to divulge. At any rate, says the tacit message, our model can match the variety proliferated by the crisis...
>
> The fact that the claim is spurious is often signalled by the hectoring attitude of the speaker, bolstered as it sometimes is by phrases such as: 'for heaven's sake; everyone knows...'; and that favorite North American version, 'aw, c'mon now'. They tell us that the regulatory model... does not enjoy requisite variety.[49]

At the level of national economies, the media play a similar cybernetic role, supporting or suppressing issues that affect people's welfare. As we can see from the work of media analysts like Robert McChesney[50] and the late André Schiffrin[51] (as well as the authors of *The Spirit Level*), the more egalitarian economies tend to have an abundance of news media, often supported from public funds, which people have the time to read and discuss; in the less equal countries media ownership is more concentrated, readership is lower, and the news tends to be dominated by celebrity stories and the scapegoating of inconvenient interest groups.

The Chilean crisis brought home violently to Beer the fact that cybernetics cuts both ways. When vested interests control the media they can use it to attenuate information they do not wish to hear, and amplify or invent the information they do.

All economies are 'knowledge economies'; the question is, how to ensure one that mediates the full variety of its people's needs, rather than using the full variety of means at its disposal to suppress them.

1. James Patrick Scanlan, *Technology, Culture, and Development*, ME Sharpe, 1992, p 44.
2. Ibid, p 44.
3. HA Simon, 'Organization and Markets'; *Journal of Economic Perspectives* 5, 2, Spring 1991.
4. Richard Stites, *Revolutionary Dreams,* Oxford University Press, 1989, p 218.
5. N Dyer-Witheford, 'Red Plenty Platforms', *Culture Machine* 14, 2013; citing figures from N Lichtenstein (ed), *Wal-Mart: The Face of Twenty-First Century Capitalism,* 2006.
6. JW Smith, *The World's Wasted Wealth*, Institute for Economic Democracy, 1994, p 234.
7. Mikhail Gorbachev, *Perestroika*, Harper Collins, 1987, p 41.
8. Michael Charles Kaser, *Soviet economics*, World University Library, 1970, p 92.
9. Stites, op cit.
10. Albert Waterston, World Bank economist, quoted by UN Economic Commission for Africa, Addis Ababa, Dec 1967, nin.tl/UNECA1967
11. Arthur Jay Klinghoffer, *Soviet Perspectives on African Socialism*, Fairleigh Dickinson Univ Press, 1969.
12. Oskar Lange, 'The Computer and the Market', in Feinstein (ed), *Socialism, Capitalism and Economic Growth*, Cambridge University Press, 1967.
13. Cornelius Castoriadis, 'Worker's Councils and the Economics of a Self-Managed Society', translated by Philadelphia Solidarity, Marxists Internet Archive, 1972, nin.tl/Castoriadis1972
14. C Castoriadis & DA Curtis, *The Castoriadis Reader*, Blackwell, 1997, p 78.
15. Jonathan Coopersmith, *The Electrification of Russia 1880-1926*, Cornell University Press, 1992.
16. Alec Nove, *Political Economy and Soviet Socialism*, George Allen & Unwin, 1979, p 42.
17. Editorial, *VESTNIK*, Journal of the Bauman Moscow State Technical University, 2005.
18. J Coopersmith, 'Soviet electrification: the roads not taken', *IEEE Technology and Society Magazine,* 12, 1993, 13-20.
19. Kaser, *op cit,* p 37.
20. John Carr, Alan Perlis, James Robertson & Norman Scott, 'A visit to computation centers in the Soviet Union', *Communications of the ACM*, 2 (6), 1959, pp 8-20.
21. From 'Silver Blaze', one of the stories in Arthur Conan Doyle, *The Memoirs of Sherlock Holmes*, 1892.

22 Alec Nove, 'The Soviet Economy: Problems and Prospects', *New Left Review* 119, 1980, p 4; cited by Geoffrey Hodgson, *Economics and Utopia*, Routledge, 1999, p 53.
23 W Paul Cockshott and Allin Cottrell, *Towards a New Socialism*, Spokesman, 1993.
24 Paul Stewart, *We Sell Our Time No More: Workers' Struggles against Lean Production in the British Car Industry*, Pluto Press, 2009..
25 'Linear Programming helps solving large multi-unit combinatorial auctions', 2001, nin.tl/linearprog ; and Juliette Garside, 'London buses show way for auction of radio wave spectrum to mobile firms', *The Guardian*, 18 Jan 2013, nin.tl/busradiowave
26 For an excellent, compact overview of the whole field, see the first chapter of William Gray, Frederick J Duhl & Nicholas D Rizzo, *General Systems Theory and Psychiatry*, Little, Brown, 1969, which is available as a free ebook at nin.tl/generalsystemstheory
27 Marc Bloch, *The historian's craft*, Manchester University Press, p 39, 1992.
28 M Dash, 'History Heroes: Marc Bloch', *Smithsonian*, 2011, nin.tl/BlochHistoryHero
29 Wikipedia entry on Jakob von Uexküll, accessed 11 March 2016.
30 Stafford Beer, *Decision and Control,* Wiley, 1966, p 34.
31 Beer, op cit, p 445.
32 Beer, op cit, p. 17 et passim
33 Mary Jo Nye, *Blackett: Physics, War, and Politics in the Twentieth Century*, Harvard University Press, 2004, p 80.
34 Ibid, p 81.
35 RW Bevan, 'Trends in operational research', *Operations Research*, 6, 441-447, 1958.
36 Enid Mumford, *Sociotechnical Systems Design*, Manchester Business School, 1985. In M Kyng, G Bjerknes & P Ehn, *Computers and Democracy*, Avebury, 1987.
37 EL Trist, *The Evolution of Socio-Technical Systems,* Ontario Quality of Working Life Centre, 1981, quoted in Kyng et al, op cit, p 73.
38 Beer, op cit, pp 482-3.
39 PN Edwards, *The Closed World*, MIT Press, 1966, nin.tl/ClosedWorld, p 116.
40 Gerald Midgley & Alejandro Ochoa-Avias (eds), *Community Operational Research,* Kluwer, 2004; Russell Ackoff, *Redesigning the Future,* Wiley, 1974.
41 Solnit, R., 2010. *A Paradise Built in Hell: The Extraordinary Communities That Arise in Disaster*, Penguin Group US.
42 Edwards, op cit.
43 Richard Barbrook, *Imaginary Futures*, Pluto, 2007.
44 A Pickering, *The cybernetic brain: sketches of another future*, University of Chicago Press, 2010.
45 From a contribution by Ashby to the 1952 Macy conference on cybernetics, quoted by Pickering, op cit, p 105.
46 W Ross Ashby, *Introduction to Cybernetics*, 1957, available at nin.tl/AshbyCybernetics

47 JFC Turner, *Housing by people*, Pantheon, 1977, available at: nin.tl/HousingbyPeople
48 Quoted by Raul Espejo from von Foerster's 'Observing systems', in Raul Espejo, 'The Footprint of Complexity', *Kybernetes* 33, no 3/4, 1 March 2004, 671-700.
49 Stafford Beer, *Brain of the Firm: the managerial cybernetics of organization* Chichester: Wiley, 1981, p 365.
50 R McChesney & JB Foster, 'The Commercial Tidal Wave'. *Monthly Review* 54, no 10, March 2003; John Nichols & Robert W McChesney, 'The Death and Life of Great American Newspapers', *The Nation*, 288, no 13, 2009, p 11.
51 André Schiffrin, *The Business of Books*, Verso, 2001; and André Schiffrin, *Words and Money*, Verso, 2010.

13

A socialist computer: Chile, 1970-1973

The overthrow of Salvador Allende's elected socialist government in Chile in September 1973 was the first act of a global neoliberal onslaught on egalitarian policies, and on the environment. Cybersyn, the Allende government's innovative, cybernetic planning project, also fell casualty to the putsch – but not before it had shown the possibility of a radically different, democratic, much lower-impact kind of technological development.

In 1979 the popular science writer Christopher Evans predicted that:

> *By the end of the next decade... another phase of the emancipation of Man from the need to work for his living will have been achieved... the seven-hour day will become a five- and perhaps even a four-hour day. And as we hand over the job of providing wealth to the computers we have created to help us, it will be further reduced and eventually the nil-hour day will arrive. The trend may have its ups and downs but its direction is inevitable.*[1]

Evans's expectations seemed reasonable at the time. He could look back on a lifetime characterized by steady progress: big reductions in working hours and improvements in working conditions; even coalminers enjoyed 'a standard of living which even the affluent middle class of Victorian times might envy'. And this improvement was echoed all over the world, including in much of ex-colonial Africa and South Asia.

Evans does not seem to have realized that the world's political direction, which for nearly 40 years had been towards greater and greater equality and autonomy, had already begun to slide into reverse.

FROM POST-WAR CONSENSUS TO SHOCK DOCTRINE

In her 2007 book *The Shock Doctrine*, the Canadian author and activist Naomi Klein dates the start of the global shift away from Utopia as 11 September 1973, in Santiago de Chile.[2] A military putsch led by General Augusto Pinochet, financed by President Richard Nixon's US State Department and by big business, overthrew the elected Socialist government of Salvador Allende, opening the way for Chile to be turned, two years later, into a test-bed for the 'Chicago school' of monetarist, free-market economics associated with Milton Friedman. Friedman believed in controlling the money supply, whatever the consequences in terms of mass unemployment, and in introducing these changes as 'shock treatment'. In the early days of the coup, Friedman even gave pep-talks to a wavering Pinochet who, although already ordering massacres and torture, was concerned about the distress that the cuts in public spending Friedman was urging would cause.[3]

The 1973 putsch in Chile, its ferocity, and the deafening silence from the 'international community' on the abuses carried out by the incoming regime, certainly sent a shockwave across the world – especially as this was quite manifestly a putsch against a democratically elected government. For the small clique around President Richard Nixon and his security adviser Henry Kissinger, the very fact that it was democratically elected made it more alarming, because this might give other electorates similar ideas, thrusting them (as the clique saw it) into the arms of the Communist bloc.[4] As Kissinger said: 'I don't see why we need to stand by and watch a country go communist because of the irresponsibility of its own people'.[5] Violent measures began even before Allende took office, as soon as it became apparent that a majority of Chileans wanted serious redistribution of wealth.

A GLOBAL CRISIS OF INEQUALITY

The Vietnam War, which would drag on until 1975, had caused massive US government debt. In 1971 the US attempted to resolve this by abandoning the post-War Bretton Woods agreement (which had maintained a degree of international financial stability) and allowing the dollar to 'float' on the international currency market.

Significant events followed one after another. Oil exporters in the Middle East raised their prices to compensate for the dollar's devaluation, then hiked them dramatically in response to US support for Israel in the Yom Kippur war of October 1973 (the month after the Pinochet putsch), raising the price of a barrel of oil from $3 to $12 in less than a year. Banks recycled the ensuing glut of petro-dollar deposits as loans to 'Third World' governments. But in the early 1980s the US, under Ronald Reagan's presidency, dramatically raised interest rates as a monetarist response to inflation. Banks, threatened with default on their Third World loans, ran to their governments. The World Bank and the International Monetary Fund (IMF) came to the banks' rescue, imposing 'structural adjustment' agreements: drastic cuts to public-sector spending and the sell-off of public-sector assets to fund debt-servicing. The burden of the crisis was borne not by the banks or even by the Third World elites who had arranged the original deals, but by the poor.

The trend intensified after the Soviet Union's breakup in 1989-90, and its subjection to what even mainstream economists now condemn as excessive shock therapy.[6] Friedmanite principles were formalized internationally in treaties such as the North American Free Trade Agreement (NAFTA, 1994) and by the creation of the World Trade Organization (WTO, 1995) which, among other things, devastated the agricultural economies of many countries by exposing them to aggressive price competition from foreign agribusiness.

Friedman himself seems to have been ambivalent about overt physical violence (he subsequently opposed US military intervention in Iraq) but his theories played well with people who had no time for equality and who had never accepted the post-War consensus, such as it was – notably Margaret Thatcher, who became UK Prime Minister in 1979, and Ronald Reagan, who became US President in 1981 (and who famously dismissed concerns about reductions in welfare support with the quip: 'some folks like to be poor').

Inequality rose all over the world. Sri Lanka, for example, had had good public-health services in 1970, and a Gini coefficient of 0.33 (where 0 is perfect equality and 1 perfect inequality), but this had risen to 0.51 by 1990 after IMF-imposed reforms. China's Gini coefficient was 0.33 in 1979 (on the eve of Deng Xiaoping's reforms) and had reached 0.57 by 1990.[7] For comparison, inequality in the UK also rose, from 0.25 in 1979 (just after Margaret Thatcher came to power) to 0.34 in 1991, reaching 0.36 by 2008[8], and was increasingly recognized as destabilizing even at this level.

Chile started the 1970s with far greater inequality than any of these examples. Chile is regarded as one of Latin America's more 'European' countries, but its Gini coefficient of 0.45 was much greater in 1970 than either Sri Lanka's or China's. After the 1973 putsch this would rise again, to match China's post-reform level of 0.57 in 1990.[9] Chile's inequality was due to two things: highly concentrated ownership of land, and control of its most profitable assets by foreign firms, so that most of the proceeds flowed overseas.

Inequality not only created the need and fed enthusiasm for Allende's program of reforms; it also fed the dynamic that drove opposition to the reforms, reflecting the complete disconnect between the worldviews of Chile's military and business elites, with their huge estates and global lifestyles, and its poor, who included thousands of landless and a large, disenfranchised indigenous (mainly Mapuche) population.

THE UNIDAD POPULAR: A MODERATELY EGALITARIAN PROGRAM

Allende's leftwing Popular Unity coalition (Unidad Popular – UP) came to power in November 1970 with what was seen as a radically egalitarian program. The actual proposals might not have seemed wildly extreme in western Europe in the immediate post-War period, but seemed thoroughly revolutionary at the time and in that place. These are the main points from the UP's 1969 manifesto:

> Basic points of government action will be: a)... to establish a system of equal minimum wages and salaries for equal work, in whatever enterprise...; b) to unify, improve and extend the system of social security... to workers who do not yet have it...; c) to assure medical and dental preventive care and treatment to all Chileans...; d) to carry out an ambitious plan of housing construction... limiting the profits of private and mixed enterprises operating in this sector...; remodelling cities and neighborhoods so as not to cast low-income people out to the outskirts...; e) There will be full civil authority [capacidad civil] for married women and equal legal status for all children whether born in or out of wedlock, as well as adequate divorce laws... with full regard for the rights of the mother and the children; f) The legal distinction between blue-collar workers [**obreros**] and white-collar workers [**empleados**] will be abolished, establishing for all the status of workers [**trabajadores**] and extending the right to organize to all those who do not already have it.[10]

The UP lacked an outright majority, but had ample support from other parties for a radical agenda for eliminating poverty and unemployment, designed to switch Chile from dependence on exports of basic commodities (especially copper) to self-sufficiency, manufacturing its own consumer and capital goods; from a colonial 'dependency' economy to an independent, modern one. This was a flat, public rejection of the official international development orthodoxy of the time: the theory of 'comparative advantage' favored by the World Bank, which, not incidentally, served the purposes of the corporations and banks of the rich countries. This maintained that poorer countries should concentrate on the things they allegedly did best, such as extracting and exporting basic commodities like copper ore, and let the more advanced countries do the more complicated, more profitable things like making cars, planes, computers and Pepsi Cola.

The UP's program depended on winning what was called 'the battle of production': the exact opposite of monetarist policy. Inflation was to be tackled, not by reducing the money supply but by increasing the supply of the goods people wanted. If this meant producing goods at a loss, so be it: effort would be concentrated on making the goods more efficiently or finding acceptable alternatives. Wages were to be increased unconditionally, and effort would be poured into identifying and then manufacturing the things people wanted, adjusting production as demand evolved.

First steps included nationalization of key industries; as Allende put it: to 'abolish the pillars propping up that minority that has always condemned our country to underdevelopment'.[11] The minimum wage was doubled immediately, and working hours cut. Results were promising: by the end of 1971 GDP was up by 7.7 per cent, factory workers' wages by 30 per cent, industrial production by 13.7 per cent, and consumption levels by 11.6 per cent.[12]

The major nationalizations were by buy-out; some were compulsory purchases under legislation inherited from the Great Depression of the 1930s. Many smaller ones, however, were unplanned: peasants and industrial workers took matters into their own hands, so that the government ended up with four times as many enterprises to manage as it had planned for.[13]

Raul Espejo, who played a central role in building the system that would attempt to transform these disparate businesses into a cybernetically co-ordinated socialist economy, wrote:

> In a very short time the government took control of more than 300 firms. This was known as the Industrial Social Area. They were the biggest firms; they represented almost 60 per cent of the country's industrial production. It was also clear that this Social Area was created to be socially responsible. There could be no question of using the laws of the market as production regulators.[14]

By the end of 1971, 68 industries, including the entire mining sector and most of the country's 23 banks, had been nationalized.

To manage all of this, Chile had a well-established but small planning and development agency, CORFO (Corporación de Fomento de la Producción) – another legacy of the progressive reforms of the 1930s. This was expanded rapidly from just 600 employees to 8,000, to cope with its new role. Not all of these new employees were in total agreement with the government's aims, but a core group of CORFO employees were well aware of the international cybernetics and Operational Research movements, and an enthusiastic handful of them knew of the British cyberneticist Stafford Beer's work. In 1971, CORFO's 28-year-old technical manager, Fernando Flores, invited Beer to come to Chile and help.

The project that emerged demonstrated, in the most demanding circumstances imaginable, a radically different way of using computers from the one that subsequently came to dominate. It was called Cybersyn in English, and Syncho in Spanish, and represented the distillation of Beer's thinking over the previous 30 years.[15]

STAFFORD BEER AND 'CYBERNETIC SOCIALISM'

By 1971 Stafford Beer was a major figure in the international cybernetics movement. He had discovered cybernetics and Operational Research while he was in the British army in India immediately after the Second World War. He read Norbert Wiener's *Cybernetics* in 1950, went immediately to visit Wiener in the US, and then launched himself on a meteoric career, setting up cybernetics departments in the British steel industry (which had just been nationalized), then an international management consultancy firm, SIGMA, and taking up a string of visiting professorships – all without the benefit of a first degree.

Beer detested the practice, already entrenched in the 1970s, of accumulating 'giant data-banks of dead information', which he called 'the biggest waste of a magnificent invention that mankind has ever perpetrated'.[16] He likened that approach to steering a car using only the rear-view mirror.

His own approach increasingly emphasized the problems created by top-down hierarchies and the need to develop the autonomous capabilities of peer groups. Like Wiener, Beer was inspired by biological systems, especially the new understanding (developed by the likes of Maurice Merleau-Ponty) of the relationship between mind and body. In this new understanding, the body is no longer the mere physical appendage of the mind, but an active and essential part of it, handling all manner of complex decisions, with barely any reference to the conscious mind. Important work on understanding the 'embodied mind' was, as it happened, being done by two Chilean neuroscientists, Humberto Maturana and his student Francisco Varela, while Beer was in Chile, and their ideas cross-fertilized with his own.

Beer's approach sought to achieve within organizations the same kind of autonomy and easy co-ordination that a healthy body enjoys. He modelled the dynamics of an organization in such a way that only small amounts of significant data needed to be gathered, and represented graphically rather than numerically, so that important relationships could be seen at a glance (for example, the gap between a factory's actual and its possible output). In an early work, he writes:

> *Very much more is learned... by inference from the system's cybernetics than from analysis of enormous masses of data... The importance of this conclusion cannot be over-emphasized. Almost the whole of government research is quite typically devoted to the collection and analysis of information about what has happened [but this is] so much flotsam on the entropic tides created by the systemic structures below the surface. Given a full understanding of those submarine structures and of the currents at depth... it becomes possible to predict effects on the surface using very little data of the former kind.*[17]

Beer was becoming more and more interested in what cybernetics could offer society, and less and less hopeful about the possibilities of doing cybernetics under capitalism. In principle, cybernetics deals with 'whole systems'. Everything that is impacted by, or which affects, the activity under consideration, in any way, needs to be brought into the model, but capitalist firms and economies think of themselves and expect to be seen as entirely self-sufficient entities, connected to the outside world only via the famous 'bottom line'. They are not happy about their other connections with the outside world being dragged into the cybernetic model, such as the ones that affect local air pollution, transport, employment, property prices, unrest in foreign places, and so on. Any attempt to consider the 'whole system' within which a capitalist firm or economy sits threatens to bring all manner of people into the equation who are not normally there, who the capitalist strongly believes have no right to be there, and whose equal presence would make it impossible to go on doing capitalism.

Eden Medina, a researcher and writer who knows Chile well, published the first book-length account of the Cybersyn project in 2011, and was able to interview many of the people involved in it at length, including Stafford Beer himself before he died in 2002. Beer told her that the timing of Flores's request was perfect. In 1971 he was just beginning his metamorphosis from Rolls Royce-driving international management consultant into a revolutionary. In late 1970 he had even delivered a lecture (to the Teilhard de Chardin Society) calling for revolution:

> Nothing else will do... We do not need to embark on the revolutionary process with bombs and fire. But we must start with a genuinely revolutionary intention: to devise wholly new methods for handling our problems.[18]

Just a few months later he received Flores's letter, inviting him to do just that. When Medina asked him, in 2001, how he felt when the invitation arrived, he replied, 'I had an orgasm!'

In contrast, the US's newly elected President, Richard Nixon, was reported to be 'beside himself' when Allende won the 1970 election, and immediately allocated $10 million to fund a coup (which failed) and 'make the economy scream'.[19] US corporations with interests in Chile were brought into the discussion – in particular, Pepsi Cola and ITT – and the owner of Chile's leading rightwing newspaper, *El Mercurio*.

As the 'crisis' of Chile's increasingly independent course

developed, the destabilization effort escalated, by overt and covert means. Overt counter-measures included cutting off US aid (reduced from $80.0 million in 1969 to $3.8 million in 1973); cutting Chile's credit rating to the lowest possible level immediately, and cutting off US bank credit altogether in 1971, so that cash had to be used for foreign transactions.

Even more severe damage was done clandestinely. An 'invisible blockade' meant greater and greater difficulties and delays in sourcing essential supplies, and in marketing Chilean goods in return for urgently needed foreign exchange. At one point Beer himself tried to help by promoting Chilean wine – at that time unknown in Europe. Firms owned by the Right were paid to go 'on strike'. Beer became engrossed in trying to fathom the cybernetics of the total situation, including the role of popular culture and song in sustaining resistance. Everything seemed to be summed up in an incident that he described later on:

> When Angel Parra wrote a haunting song called 'El Barque Phantasmo' *about the ghost ship carrying copper which no-one would unload, and when Allende approved it to the extent that he wished to take disc pressings as gifts to the members of the United Nations he was shortly to address, the President was thwarted. The record-pressing company was on strike.*[20]

Over the next three years US government and corporate agencies funneled lavish funds to Chile's rightwing media, to far-right terror groups, to finance murders and sabotage, and even to pay people to exacerbate shortages by buying up sugar and dumping it in rivers.[21] None of this succeeded in bringing down the Allende government – which instead won greatly increased electoral support in the elections of May 1973. The Cybersyn project played a small but important role in Chile's annoying failure to succumb.

Cybersyn was designed to help manage the economy in real time and identify problems ahead of time. It would give a complete and up-to-date picture of what was happening in the economy, in as much detail as necessary for ensuring supplies of what was needed, but no more than that: how much capacity was being used, how much was available, of what kinds, and where. This fitted naturally with the 'battle of production' approach to managing the economy, and although the underlying cybernetic principles were challenging, the physical implementation did not need to be complicated.

As Eden Medina says, 'it is important to note that Cybersyn engineers were not interested in financial information: [they were] focused exclusively on industrial production'. New but simpler criteria could therefore be used (such as current, potential and theoretically achievable output levels) which could, moreover, be expressed as 'unitless percentages' or 'quantified flow-charts'.[22]

The intention was to support the autonomy of workers and work groups, and to strip away the kind of 'dead data' that accumulate in typical computer systems. In his report to the Chilean government in January 1972, Beer wrote:

> It is a primary aim to avoid creating a vast bureaucratic machine, and the true intention... is simply to discard all the data once they have been wrung dry by this powerful online system.[23]

HOW MUCH COMPUTER HARDWARE DOES A VIABLE SOCIETY NEED?

Insofar as Cybersyn has attracted public interest, it has tended to focus on the physical computer that was used in the project. At the time, this even took the form of scare stories that Chile was being managed by a sinister 'electronic brain' – including a very unhelpful article that was published in Britain, in *The Observer*.[24] In fact, the computer itself played very little part in the project's very impressive achievements because it could not be fully deployed in time. It was in any case a very modest computer, although its software was advanced, involving a new and sophisticated system of statistical filtering (Bayesian filtering, used nowadays in email spam-filters, among other places) to distinguish significant trends that might otherwise escape notice and discard insignificant information, yielding a sparse stream of highly relevant, useful and above all timely information.[25]

It was just as well that Cybersyn wasn't planned as a massive data-capture, data-mining exercise: there was not enough computing capacity in all of Chile even to contemplate such a thing. Ideally, the system would have provided a computer for each factory, shop and depot, to give its staff a full picture of their own outstanding orders and available resources, in real time, and warn them of possible problems without reference to any higher authority – an ideal task for the cheap, personal and home computers that were becoming plentiful 10 years later.

The government had only four mainframe computers – none of which worked in real time (as a modern computer does, allowing you to enter your data and see the results immediately). Cybersyn was given full use of one of them, a Burroughs 3500: a somewhat eccentric 1966 design with 500k of main memory. This somehow had to be made to simulate a computer network consisting of hundreds and potentially thousands of nodes (in the factories and utilities). It looked, and physically was, a top-down affair, and this (and the exciting-looking command-center, the Opsroom, unveiled later) helped to foster the perception of a 'big brother' system, running the economy by remote control – something that Cybersyn had nowhere near the capacity to do.

In lieu of real-time processing, information had to be gathered from the workplaces each day, run through the machine in 'batch mode', and the results passed back to the workplaces the next day.

There was also the problem of how physically to connect the workplaces to the machine. A cheap but effective method was hit upon: to connect the enterprises by telex machines, which could operate over telephone lines and a network of microwave links that had been installed in the 1960s for satellite tracking. Even this was not straightforward: there were not enough telex machines and more could not be bought because Chile was now under unofficial but very effective blockade. But then a CORFO employee discovered a forgotten hoard of 400 brand-new machines, ordered in the 1960s but never used, sitting in a store-room at ENTEL, the National Telecommunications Enterprise.

The telex machines weren't even directly connected to the computer at first – all data had to be re-entered in Santiago –

but this was never a major problem as there wasn't too much of it.

The revolutionary aspect of the network wasn't the technology but the type of data being captured, and its form: essentially just three figures for each of the key parameters affecting any given business (parameters which might include machine availability, stock-levels or staff availability). These three figures were 'actuality', 'capability' and 'potentiality', of which the first two might represent current output and spare capacity, and the third, 'potentiality', a calculation of what might be achieved under ideal circumstances. Calculating and checking these figures, and teaching factory staff how to use them, took a large part of the project's effort, but, once defined, the system could derive a surprising amount of useful information from them, and they could ultimately be represented iconically, like the 'levels' display on a hi-fi system. Beer had developed the principle in his book, *Brain of the Firm*, which he had just finished before he went to Chile, and which became the project's 'bible'.[26]

Cybersyn's most famous feature was a prototype 'Opsroom', which looked like something from a James Bond movie or the TV series *Star Trek*. Again, there should ideally have been hundreds of these, in workplaces up and down the country. Planners, political and labor representatives, seated in big, comfortable swivel-chairs with control panels in the armrests, could summon up graphic displays on the surrounding walls illustrating every aspect of the country's economic life, using data no more than a few hours old. Using the armrest controls, they could explore the data 'as if using a hypertext', according to one of Medina's interviewees – a full 20 years before that kind of experience became commonly available with the World Wide Web.

These displays did sophisticated things, yet used what might sound like unsophisticated means: a system of slide-projectors, controlled by skilled assistants who assembled each screen as it was requested from a 'vocabulary' of specially designed iconic images on slides. This was intended to be a complete graphic vocabulary covering all foreseeable possibilities, and there seem not to have been any plans or desire to automate the displays. This was intended to be a social system, not just a technical one.

Two temporary control rooms were built and brought into service in 1973, not as sophisticated as the prototype Opsroom, but with much the same functionality, which was conclusively demonstrated, however; so much so that Allende considered having the main Opsroom moved into the presidential palace in the days preceding the final coup.

THE OCTOBER 'STRIKE' EXPOSES THE REDUNDANCY OF CAPITAL

Even before it approached completion, the system helped to frustrate two waves of 'capital strikes' organized by rightwing groups and the small business organizations known in Chile as '*gremios*' – often mistranslated at the time as 'unions' by the foreign press.

The October 'strike' was not a strike in the normal sense but more like a lockout. It started as a protest by private truck owners against the creation of a state trucking service in the Aysén region of southern Chile; it spread with enormous rapidity and it became clear that there had been widespread preparation for it. There were road-blocks; stores closed, and 'enforcement squads' attacked stores and vehicles that refused to join the strike. The National Manufacturers' Association locked out their employees, in some cases paying them to stay at home. Medina describes it as 'a public demonstration of class power by the bourgeoisie'. It later emerged that the strike was lavishly funded by the US State Department and business.[27]

This very determined and well-financed effort to overthrow the government was defeated by the end of October; partly by the sheer energy of Chilean workers and community organizations, who occupied factories and shops, defended them, and organized food distribution direct to people who needed it; and partly by Cybersyn. Very few people were aware of the project at that time, and many of those who did witness its work in beating the strike did not realize what they were seeing: all they saw was a roomful of telex machines thundering away. ('The noise was deafening', Beer wrote.)

The attempt to shut down the economy in October 1972 not only failed; it also shone a floodlight on the sheer scale of the redundancy (in terms of unnecessary equipment and so on) a supposedly 'efficient', competition-based economy needs, just so that its players can compete with each other.

Normally, servicing Chile's economy required 40,000 independently operated trucks; during the 'strike', no more than a third of that number did the same job, and often just a tenth or even fewer. According to Gustavo Silva, one of the civil servants involved, the network made it possible for just 200 trucks to do enough of the work normally done by the national fleet of 40,000 to keep the economy moving.[28]

He told Medina: 'We knew exactly how many trucks we needed, so each time we lost one we were able to requisition another'... 'Two concepts stayed in our minds: that information helps you make decisions and, above all, that it helps you keep a record of this information... correct your mistakes and see why things happened.' For example, they might get a call saying that there was no kerosene in such-and-such an area. 'But why? We sent a truck there!' In minutes, they could find out what had happened. In this way, says Medina, the network connected the 'revolution from above' with the one from below.

In his own account, Beer explains the deeper implication of the victory: redundancy of vehicles and physical resources had been replaced by 'redundancy of control'.

> *The huge surge of information into [the network] operated as a negentropy pump: instant communication loops sprang into being, and instant decisions were available... [this] redundancy of potential command is decentralizing, and it is robust.*[29]

In other words, the country's normal 'knowledge economy' was turned upside down: instead of the owners of resources hoarding knowledge for their own purposes, all of the significant knowledge was suddenly in the hands of the planners. They could see what needed to be done and could generally choose from a variety of responses, even with drastically fewer resources. For once, it was

A socialist computer

the owners of capital who were in the dark.

Raul Espejo, who worked as project manager throughout, wrote a brief account of it in November 1973, after his arrival in England. In this he was at pains to emphasize the human aspect of the system; many people, he wrote, got very excited about the technology – but this was not the main thing. The computer system was not complete and could not, in any case, have provided the real-time analysis the situation demanded. In October 1972 it was the telex network, and the people using it, that saved the day: the computer's sophisticated Bayesian filters played no part. Espejo wrote:

> The organizational model which we were building scientifically, by now existed unconsciously in the minds of managers; and now the new information routines we had been instituting suddenly acquired a meaning and the communication network a physical reality. The correlation between information and decision, which was missing in stage 2, suddenly became easy... People immediately found that they got instant answers or decisions as soon as their problems were raised.
>
> The comprehension of this fact exploded like a bomb on the minds of all concerned.[30]

The experience of defeating this attempted counter-revolution gave a great boost to the radicalization and self-organization on the streets and in the factories. Fifty more factories had been taken over by their workers by the end of the strike; only 15 of them were returned to their owners afterwards.

Work on Cybersyn was galvanized, and momentum gathered through the remaining 11 months of Chile's democracy. In November 1972, the software for the national planning system started producing its first daily printouts, checking a growing range of production indicators for anomalies. The Opsroom was functioning by the following January,[31] and it worked – although sometimes in ways that showed up frustrating shortcomings in the reporting system. Medina records an occasion when it correctly predicted a coal shortage at a cement works, several days *after* the crisis had passed: the plant managers had delayed sending in their daily capability figures, having not understood their significance.[32]

A second strike, in August 1973, was defeated with even greater ease. Espejo recorded from notes made at the time that although 'no more than 10 per cent of the truck fleet was working' at the beginning of the strike, and no more than 30 per cent by its end, fuel supplies remained 'virtually normal', as did the flow of raw materials to the strategic productive sector, and food supplies remained 'perfectly adequate' throughout the country.

CHEAP, RADICAL TECHNOLOGY

Beer became increasingly radicalized and expanded his vision for the system. He looked for ways to apply cybernetic thinking to counteract the power of the Chilean media, which were nearly all owned by the Right, and to bring workers more into the decision process.

One initiative led Beer to make contact with popular musicians and singers in the Nueva Canción movement – of which Victor Jara was a leading figure: the famous political singer subsequently tortured and murdered by police in the National Stadium after the September putsch. Beer evaluated their contribution to the revolution in engineering terms – as an information channel that might counteract the flood of anti-government rhetoric coming from the mainstream media.

Another project, known as 'Cyberfolk', included a simple system that could be used, for example, at meetings, where employees might feel inhibited about expressing their candid opinions. It was built in England by Beer's son Simon, an electronics engineer, and consisted essentially of 'no more than one loop of low-tension wire'.[33] Unlike the voting machines that appeared subsequently in Western TV programs, where the audience is asked to vote on specific questions at specific times, this one allowed for continuous and variable-strength feedback throughout the discussion. Dials, connected by the wire, allowed employees discreetly to register their levels of enthusiasm for proposals, as the discussion progressed, the collective level of approval being displayed on a 'summation meter', visible to everyone. Since the boss, or speaker, could only see 'the feeling of the meeting', no individual need fear reprisals for lack of enthusiasm.

The Cybersyn team tested the prototype in meetings at the Catholic University in Santiago and it 'worked very well', according to Raul Espejo.[34] According to a recently created Chilean web archive about Cybersyn, the system was tested with cable TV in two cities, Tome and Mejillones.[35] The ultimate idea was to use this as part of a 'people's TV channel' to mediate a live, national policy dialog with the electorate but, as the archive's authors point out:

> *This system required total commitment and honesty from both parties, a situation that was impossible in those times owing to the socio-economic instability that characterized the country. Those groups that opposed the Allende Government wrongly accused this experiment as a tool to control people, when in all truth it would be the people who would finally have the chance to take part in the civic decisions of the State.*

'WAR' IS DECLARED

All of this was unbearable provocation to the Right. Their violence increased and became more and more overt. Allende refused to distribute arms to the workers, despite being urged to do so by Cuba's president, Fidel Castro, the minister for the economy Pedro Vuskovic, and many others; some workers' groups armed themselves anyway, but when the putsch came, resistance was essentially non-existent, and the Right had to use their imaginations when identifying targets for attack.

On 11 September 1973, Pinochet sent jet fighters and tanks against the undefended Moneda presidential palace. Allende carried through his promise to leave office only 'in wooden pajamas': his coffin. He shot himself, probably with the gold-plated Kalashnikov rifle that Castro had given him, to avoid being paraded as a trophy by the junta. Pinochet and his backers replaced the elected government with a system of murder, torture and 'disappearance', whose thousands of victims are still not all accounted for more than 40 years later.

The economy slumped in 1975, was subjected to full monetarist discipline, and collapsed completely in 1982. It then recovered – partly thanks to decisions to retain the lucrative copper industry in the public sector (source of around 30 per cent of government income) and to sell large areas of land to foreign agribusiness. Foreign aid was resumed, including World Bank loans, and Chile became, officially, a success story, to the extent that people of distant lands can now buy Chilean apples instead of locally grown ones, from Dole, Del Monte and Fyffes.[36] It still has the highest inequality of any OECD country (with a Gini coefficient of 0.52) and 38 per cent of its citizens find it 'difficult or very difficult to live on their present income', but at least they have been promised that Chile will become a 'developed country' by 2018.[37]

The Chilean intervention was intended to tell the people of the world what would happen to them should they take ideas of freedom, democracy and equality too seriously. But it was also an admission that social justice was not pie in the sky but a practical possibility that, in the end, could only be prevented by brute force – an admission that so violated even the moral sensibilities of Pinochet and his forces, that they flung themselves heart and soul into delusion, attacking empty streets with real tanks, jet fighters and rockets. Till his dying day, Pinochet referred to his putsch as a 'war'.

Chile's 'first 9/11' may have inaugurated the era of 'disaster capitalism' but the final resort to undisguised lethal force also revealed to the world that the polite doctrines that for centuries had given a veil of legitimacy to the domination of the weak by the powerful were bankrupt. The wheels had fallen off all their trusty old rationales one after another: the superior wisdom of markets; the stupidity of the working class; eugenics; the 'civilizing mission' of the better-armed nations... all that remained was brute force wrapped up in self-delusional histrionics.

Some months later, in a radio lecture in Canada, Beer said: 'I tell you solemnly that in Chile the whole of humanity has taken a beating'.³⁸ He added a substantial account of the project to the 1981 second edition of his book *Brain of the Firm* and, in this, he wrote:

> I have often been asked why we were not able to stipulate a behavior which would accommodate that threat [of external intervention]. It is like complaining that man, who is supposed to be an adaptive biological system, cannot adapt to a bullet through the heart.³⁹

The environment also took a beating as a result of the Pinochet putsch. Allende's programs for land redistribution and development of the rural economy were of course scrapped, replaced by an unchallenged handover of natural resources to the extractive industries. Chile lost two million hectares of its unique native forests in the single decade from 1985 to 1995; 80 per cent of what's left is now in private hands. In 1999, when Miguel Altieri and Alejandro Rojas surveyed Chile's ecological situation in a journal article, central Chile had lost all of its native forest and the fragile cloud-forests of Tierra del Fuego were being targeted by the wood-pulp industry. Fisheries were also suffering extreme and probably irreversible over-exploitation; agricultural land was suffering some of the world's worst levels of soil-loss, the damage being compounded by escalating use of artificial fertilizers and pesticides. Altieri and Rojas wrote: 'It could be argued that throughout the authoritarian regime (1973-89), not only was there no environmental policy, but its absence was also considered an advantage in attracting foreign capital to Chile'.⁴⁰

This was not just Chile's tragedy. The myth of Chile's economic success was used to promote and enforce neoliberal policies worldwide during the 1970s, 1980s and 1990s, adding major impetus to what the ecologist Barry Commoner called the 'Great Acceleration' of global ecological impacts.⁴¹

The word 'inevitable' has often been applied to the fall of Chile's socialist government in 1973. Jonathan Haslam's account, which is generally sympathetic to Allende, is subtitled 'A case of assisted suicide'. Alec Nove (the socialist economist whose downbeat views on wealth redistribution are mentioned in Chapter 4) was in Chile during the Allende period and argued afterwards that the project had been doomed because it 'alienated the petit-bourgeoisie'. He heard with his own ears 'housewives banging saucepan lids nightly in protest at food shortages'.[42] But these vociferous protests, by better-off minorities, played very little part in the government's overthrow, which, unknown to Nove, was being plotted by much bigger forces.

Allende's overthrow was no more inevitable than the somewhat freakish combination of factors that made it possible. It was not inevitable that an informal clique would be controlling US foreign policy in 1970, occasionally sidelining even the CIA. Or that Nixon would survive the revelations about his other criminal activities until 1974; or that he would even have become US President. Nor was it inevitable that the Vietnam War would still be distorting US politics in 1973, in the teeth of a decade of escalating opposition, and especially after Daniel Ellsberg's 1971 exposure (in the 'Pentagon Papers') of the criminality behind it. Or even that such a thing as a 'Vietnam War' would have been conceivable after 1945, let alone one involving the US, when it seemed that the people of Vietnam had already earned their independence from France during the Second World War, in their successful campaigns against Japanese occupation.

These defeats of democratic process were certainly possible, or somewhat likely, or perhaps even highly probable, but none of them was pre-ordained by fate. Every single step on the road to September 1973 was contested – and events could have taken a completely different course. The same is equally true of campaigns for equality and justice, many of which turned out to be possible after all, despite loud insistence from all quarters that they were not: the battles against the Atlantic slave trade; for women's emancipation; against capital punishment in many countries; for gay rights; and a succession of dictators, tyrants and crooked politicians have been overthrown, exposed, ejected from office, and sometimes even jailed or executed. There is clearly no fundamental law that says brutality rather than humanity will prevail.

Progressive initiatives are always doomed to fail, until they don't, and the same is true of oppressive ones.

A socialist computer

1. Christopher Riche Evans, *The Mighty Micro*, Gollancz, 1979, pp 150-1.
2. Naomi Klein, *The Shock Doctrine*, Metropolitan Books/Henry Holt, 2007.
3. Ibid, pp 80-81.
4. Jonathan Haslam, *The Nixon Administration and the Death of Allende's Chile*, Verso, 2005, pp 55-58.
5. William Blum, *Killing Hope*, Zed Books, 2003, p 209.
6. Joseph E Stiglitz, *Globalization and Its Discontents*, WW Norton, 2002.
7. Donald Lamberton, *Managing the Global*, IBTauris, 2002, p 182.
8. Jonathan Cribb, *Income inequality in the UK*, Institute for Fiscal Studies, Feb 2013.
9. Lamberton, op cit, p 182.
10. From the *Programa UP*, 1969, quoted in V Wallace (translator), 'The Notion of Equality in Chile's Communist and Socialist Left, 1960-1973', *Socialism and Democracy*, 27(2), 2014, nin.tl/equalityChile
11. Allende, quoted in Eden Medina, *Cybernetic Revolutionaries: Technology and Politics in Allende's Chile*, MIT Press, 2011, p 15.
12. Medina, 2011, op cit, pp 50-52.
13. Ibid, p 52.
14. Raul Espejo, 'Cybernetic Praxis In Government: The Role Of The Communication Network', unpublished MS, Nov 1973.
15. Stafford Beer added a very full account of the episode to the second edition of his book *Brain of the Firm*, Wiley, 1981. Eden Medina has produced the current, definitive account in *Cybernetic Revolutionaries*, op cit.
16. Stafford Beer, *Platform for change : a message from Stafford Beer*, Wiley, 1975, p 431.
17. Stafford Beer, *Decision and control : the meaning of operational research and management*, Wiley, 1966, p 479.
18. Platform for Change, op cit, p 36.
19. Haslam, op cit, 2005, p 67.
20. Beer, *Brain of the Firm*, op cit, p 308.
21. Reported to Eden Medina by Gustavo Silva; Medina, op cit, p 164.
22. Medina, 2011, op cit, p 129.
23. Beer, *Brain of the Firm*, op cit, p 263.
24. Medina, 2011, op cit, p 173.
25. Beer, *Brain of the Firm*, 1981 op cit, pp 262-3.
26. Beer, *Brain of the Firm*, 1981 op cit, p 164.
27. Medina, 2011, op cit, p 146.
28. Eden Medina, 'Designing Freedom, Regulating a Nation: Socialist Cybernetics in Allende's Chile', *Journal of Latin American Studies*, 38 (3), 2006, pp 571-606.
29. Beer, *Brain of the Firm*, op cit, p 313-4.
30. Espejo, op cit.
31. Medina 2011, op cit, p 165.
32. Medina 2011, op cit, p 198.
33. Beer, *Brain of the Firm*, op cit, p 285.
34. Interview with the author, Jan 2011.
35. Enrique Rivera, Catalina Ossa, Daniel Opazo, Sebastián Vidal, Benjamín Marambio & Edgard Berendsen, *Cybersyn/Cybernetic Synergy*,

nin.tl/1XfCAm2 Accessed 4 July 2015.
36 No author given, 'Chiquita Brands International Inc' nin.tl/ChiquitaDuke Accessed 25 Sep 2014. See also Philip McMichael, *Food and Agrarian Orders in the World-Economy*, Greenwood Publishing Group, 1995, p 134.
37 COHA, 'The Inequality Behind Chile's Prosperity', nin.tl/Chileinequality Accessed 24 Sep 2014.
38 Stafford Beer, *Designing Freedom*, Canadian Broadcasting Corporation Massey Lectures, 1973. Text available at nin.tl/BeerFreedom ; audio available at: nin.tl/BeerFreedomaudio.
39 Beer, *Brain of the Firm*, op cit, p 346.
40 MA Altieri & A Rojas, 'Ecological Impacts of Chile's Neoliberal Policies, with special emphasis on agroecosystems', *Environment, Development and Sustainability*, 1, 1999, pp 55-72.
41 Barry Commoner, *The Closing Circle*, Alfred E Knopf, 1971.
42 Alec Nove, *The Economics of Feasible Socialism*, George Allen and Unwin, 1983, p 161.

14

Utopia or bust

If we are to save the planet, it will mean developing a vivid sense of the kind of world we want – and that may involve taking the whole notion of Utopia more seriously. The conventional assumption that humans are fundamentally selfish and destructive is not supported by evidence – and is actively disproved by the sense of community and caring displayed in the wake of disasters. In a more equal world, jobs would serve community needs rather than profit, caring roles would be a priority and automation would encourage skilled work rather than eliminate it. But to arrive there we will need to undermine the 'apparatus of justification' on which inequality depends.

The global environmental crisis had been apparent before 1973 but in or around that year a tipping point was crossed: the global economy finally moved into 'eco-deficit'. It started using biological resources faster than natural processes could regenerate them, so that (as the Global Footprint Network think tank put it) 'by the late 2030s humanity will need the equivalent of two Earths to keep up with our demands'.[1] At the same time, climate change began to become a public and political issue. James Lovelock spelled out the implications in his 'Gaia Hypothesis' in the 1970s, publishing it as a best-selling book in 1979.[2]

Over the same period, the diversity of plant life has collapsed, as farming has become more and more industrialized.[3] A WWF report published in December 2014 found that the world had lost half of all of its vertebrate life forms in the four decades from 1974; most of this loss was due to intensified exploitation of habitats for commercial purposes.[4]

These times demand radical action yet a tragic fatalism has paralyzed our politics. This is not helped by the cod-Darwinian notion of 'progress' that makes it difficult to imagine alternatives. We are encouraged to believe that a 'technological revolution', which few of us can understand, still less shape to our needs, is following its predestined course. We seem rather like Hollywood extras in this revolution. It will happen anyway, with us or without us, the path being blazed for us by a supposedly brilliant few, without democratic assistance.

In 2013, the New Yorker's George Packer found that companies like Google and Facebook are full of people who fervently believe they are changing the world more effectively than any government can, and that it is entirely appropriate to become extremely rich by doing so. Packer found the phrase 'change the world' used constantly in these companies and among their backers, yet they were surrounded by (and oblivious to) levels of homelessness and poverty that had been unknown in San Francisco a couple of decades earlier.[5]

The Chilean experience of 1970-73 provides a reality-check to the revolution we've been offered. Cybersyn was there to support a drive for equality; for the production of the things people needed and wanted in the here-and-now rather than in some hypothetical future, within an economy organized for that purpose. This was revolution in the real world, focusing on real-world practicalities, including land redistribution and a turnaround in the economy from dependency to self-sufficiency, producing even the kinds of consumer goods 'developing countries' were supposed to buy from wealthier ones. This was done by public, political decisions and debate, and by individual and collective action. For some Chileans this was an outrageous violation of fundamental laws of the universe. For many, it was a real taste of Utopia.[6]

Then, we still have the ingrained belief, alluded to in the last chapter, that oppression is somehow destined to prevail and that humane projects, like Chile's democratic socialist revolution, are doomed to bitter failure. The 17th-century philosopher Thomas Hobbes's contention that humans are fundamentally selfish and destructive, and therefore need strong rulers, retains a powerful hold. Objective evidence increasingly discredits the Hobbesian version, but that doesn't stop rightwing ideologues from using every scrap of evidence they can lay their hands on to 'prove' that we aren't as nice as we think we are – and it doesn't even prevent many apparently democratic and liberal types from going along with it.

A spectacular example of gratuitous Hobbesianism appeared in a column by the leftish-leaning Oxford historian and columnist Timothy Garton-Ash, after the city of New Orleans was hit by Hurricane Katrina in 2005, killing nearly 2,000 people. Instead of addressing the disaster's implications for the global-warming debate, or condemning the sheer economic injustice that had starved New Orleans's flood-protection system of necessary funds and left thousands of people to fend for themselves for days on highway overpasses and in buildings without electricity or clean water, he wrote a loud lament against human nature, inspired by early rumors of looting and mayhem that he had picked up from the media:

> *Katrina's big lesson is that the crust of civilization on which we tread is always wafer thin. One tremor, and you've fallen through, scratching and gouging for your life like a wild dog... remove the elementary staples of organized civilized life... and we go back within hours to a Hobbesian state of nature, a war of all against all.*[7]

It turned out that rightwing media and think tanks had systematically concocted enormous lies that bore no relation whatever to actual events.[8] When the true story finally emerged it received little coverage. As we will see below, Katrina's 'big lesson' was about patience, heroism and mutual aid on a huge scale, but that did not fit with the media or government narrative, or even with the deep assumptions of some quite liberal-minded people.

Why this readiness to accept the Hobbesian counsel of despair? Could it be that despair is a kindlier bedfellow than the nightmare of facing and challenging the overwhelming terror that modern governments can unleash? Does it go: 'I can see the situation is unjust but, if humanity is fundamentally flawed, then there is no point struggling for a better world and I am off the hook'?

ENVISIONING UTOPIA: THE WORLD TURNED RIGHT WAY UP

After the 2007/8 financial crash expectations were high of a global, popular backlash against the system that had created the crisis, but the system has survived. In 2015, the rich were getting richer than ever and the poor, poorer. By January 2016 just 62 people controlled as much wealth as the world's poorest 50 per cent, some 3.6 billion people.[9] The environmental crisis deepens commensurately, and the forces that are causing it proceed almost unhindered.

Why so little opposition to them, and so muted? Could it be that we lack a positive and lively sense of the kind of world we want?

The case for serious political change is made in generally negative terms. 'The usual green promise,' says environmental writer George Monbiot, is 'follow us and the world will be slightly less crap than it would otherwise have been.'[10] The message is often some version of 'unless we change our ways of thinking about consumption and resource use, unless we demand that government provide really sustainable solutions, we as a society and species are headed for disaster',[11] which is true but bleak. A 2013 statement called 'Premises for a New Economy', drafted by a group of 18 sustainability experts led by the economist Stephen Marglin, called for 'an actual reduction in conventional measures of standard of living', but also indicated the need for

> an economics that places higher value on discretionary time [that], in part, would supplant private consumption with new public amenities and spaces that create non-commodified opportunities for leisure and self-development. A second substitution is to build community and other forms of human connection, thereby enriching people's lives without enlarging ecological footprints.[12]

...which sounds potentially exciting, but is tantalizingly vague.

We need a more detailed vision, and a sense of how it might be reached, starting from where we are now. The technological stories described in the earlier chapters show, I hope, that there are always plenty of alternative paths for human development, and how tolerating inequality means that those paths become closed off to us as positional interests close in. We can begin to develop a sense of how the size and power of those positional interests, and the presence or absence of constraints on them, predict the speed and thoroughness with which alternatives are eliminated and obliterated, and why the most harmful alternatives will tend to be favored when power inequalities are allowed free play.

Rolling the film backwards, as it were, we can even begin to develop a reasonably detailed idea of what those alternative paths might be, and even of what life would look and feel like subjectively, if we were to grasp the nettle, and call time on inequality.

UTOPIAN PRACTICALITIES: FOOD AND WORK

Food is a good indicator of what happens to diversity, health, and human environmental impact as the battle between elitism and egalitarianism ebbs and flows. As inequality advances (for example, through the free-trade deals that oblige countries like Mexico to accept subsidized US food imports, or which favor the eviction of peasants from their land and its sale to foreign agribusiness) food diversity collapses and environmental impact soars. The industrial agriculture that's encouraged by this process produces a large part (about 14 per cent) of global greenhouse-gas emissions; a further 17 per cent are added by deforestation, which is largely for the benefit of large-scale, commodity crop and meat production. Yet, for all its impact, this kind of agriculture produces only 30 per cent of the world's food.[13]

When the contest goes the other way, and peasant movements reclaim land from big estates and farm it for food rather than profit, diversity tends to return and the soil becomes an absorber rather than an emitter of greenhouse gases (largely because peasants and smallholders tend to farm without artificial inputs).

Via Campesina, the worldwide peasant movement for food sovereignty founded in Brazil in 1993, helps to co-ordinate and has documented many inspiring examples of this counter-movement from all over the world. There are some particularly inspiring examples from Zimbabwe and Tanzania, for example, where tens of thousands of peasants are implementing organic methods and restoring land that had been depleted during colonialization.[14]

Throughout modern history, peasant agriculture has been portrayed as the epitome of backwardness, and treated merely as an easy source of revenue and cheap human labor. It has almost never been on the receiving end of social investment, let alone on the scale that investment has been showered on market-oriented industrial agriculture. Even so, it still produces 50 per cent of the world's food. A further 25 per cent is now produced by small-scale agriculture within cities. Peasant and small-scale agriculture also provides the rich diversity of food that industrial agriculture has abandoned. And where this kind of farming receives even quite modest investment, huge increases in yield are often achieved – using entirely organic methods. In Nepal, increases of 175 per cent are recorded; in Tigray, Ethiopia, yields have been reported that are three to five times greater than those achieved using chemical fertilizers.[15]

IS YOUR JOB REALLY NECESSARY?

We are encouraged to think that being modern means being busy – but, as Chapter 8 argued, when you strip away all the sales-oriented effort involved in a modern economy, not a lot is left.

We might all work a lot less in Utopia than we do now, but it depends on what you call work. In the early 1900s a North American agronomist, FH King, who studied peasant agriculture in China, remarked on the prevalent notion in the US that Chinese people were lazy. Yet wherever he went in China, he found extraordinary productivity and creativity. The apparent contradiction stemmed from the peasants' different approach to work – more like art than like toil:

> *The oriental farmer is a time economizer beyond all others. He uses the first and last minute and all that are between. The foreigner accuses the Chinaman of being always long on time, never in a fret, never in a hurry. This is quite true and made possible for the simple reason that they are a people who definitely set their faces toward the future and lead time by the forelock.*[16]

His interviewees produced enormous amounts of food from tiny plots of land without any artificial inputs and not a scrap of waste, generally managing to live fairly well despite extortionate rents. Without those rents their lives really would have been ones of ease. The key to the riddle of the 'lazy' Chinese peasants and their long, creative working hours clearly lay in their famous culture of reciprocity, known as *guanxi*.[17] These people were oppressed, but at the village level there was great equality and solidarity, and from this came an autonomous approach to work, setting their own schedules and tasks, and the overall sense of good-humored self-assurance and strength that King fell in love with.

Similar things have been written about traditional industries in England. George Sturt, in his classic *The Wheelwright's Shop*, wrote that:

> *...in those days a man's work, though more laborious to his muscles, was not nearly so exhausting yet tedious as machinery and 'speeding up' have since made it for his mind and temper. 'Eight hours' today is less interesting and probably more toilsome than 'twelve hours' then.*[18]

A 'chair-bodger' from Herefordshire called Phil Clissett (who taught the art to architect and craftworker Ernest Gimson in 1890)

> *...could turn out his work from cleft ash poles on his pole lathe, steam, bend and all the rest. He seems to have made a chair a day for 6/6d [six shillings and sixpence] and rushed it in his cottage kitchen singing as he worked. According to old Philip Clissett, if you were not singing you were not happy.*[19]

THE BLEEDING EDGE

A lot of the most productive work that people do looks like playing around or staring into space. Some of the most counter-productive, or even destructive work is done in paid employment. And the world of paid employment barely acknowledges or actively impedes the most important work of all: caring work, especially looking after little children and old people. Hunter gatherers offer an interesting reality check. Most of them live in harsh, marginal environments where you would expect survival to be very hard work indeed. But Richard Lee measured the working hours of the Dobe indigenous group in 1969 and found that they spent on average 2.5 days a week 'working' (much of the time spent gathering and hunting was heavily diluted by breaks for conversation, sleep and so on). Even lower working hours were recorded among people in Arnhem Land, northern Australia, and workloads were similar or not much greater all over the hunter-gatherer world, from Lapland to Tierra del Fuego.[20]

Can an industrial society offer lives of similar ease? Otherwise, what's the point of living in one? In 1905, the novelist HG Wells thought that, in view of the levels of automation then available, there need be 'no appreciable toil in the world'.[21] In 1930, the economist John Maynard Keynes reckoned that a 15-hour week would be possible by 2030.[22] In 1977 the Adret collective, a group of French workers that included a docker, a nuclear physicist, a secretary and a factory shift-worker, put their heads together and published their assessment of the possibilities under the title *Travailler deux heures par jour*[23] ('Working two hours a day'). Two years later Christopher Evans (see Chapter 13) was predicting a zero-hour day.

There is a catch to predictions like these. They are often based on an assumption that we are comparing like with like when we compare the time taken to make the goods we need now and in the past. An automobile, or a shirt, or a loaf of bread that's produced so quickly now is a very different thing from the automobile, shirt or loaf of 20 or 50 years ago – sometimes in good ways, but also in regrettable ways. Automation under capitalism is less to relieve drudgery than to relieve manufacturers of some of their wage bills and reduce their reliance on skilled workers.

The work content that has been squeezed out of automobiles, for example, includes the hand-stitched upholstery and lacquered paint finishes and many other nice touches that collectors of veteran cars like so much. The work content that has been squeezed out of the food chain has removed locally grown food from most people's diets and many varieties of fruit and vegetables, often species with higher nutritional value, because they needed more labor to pick and process.[24] The work content that has been squeezed out of houses has left them without stained glass, ornamental plasterwork, moldings and tile work, the clever paint finishes that Robert Tressell's 'ragged trousered philanthropists' knew how to do,[25] and panelled doors.

Much of the work that has been squeezed out of the process is work people enjoy doing, and the only thing wrong with it was that it was so badly paid. And people tend strongly to prefer the things it produced. The rich always make sure they have access to it.

A hallmark of egalitarian societies is that skill is a major social asset and source of personal value, and this is exactly what makes the villages and clothes and artefacts of traditional societies so very attractive to Westerners. If we could eliminate or merely reduce wealth inequality we could find human work flowing back into our material environment as the positional forces that have concentrated workplaces into larger, more widely separated units, lose their power and importance. It would also be promoted actively, through the kinds of economic policies more egalitarian countries tend to pursue, almost irrespective of their official politics. Cuba and the Scandinavian countries have successfully pursued educational policies aimed at providing the entire range of options to all children, even in the remotest parts of the country.

After the Second World War Norway adopted a policy of protecting and supporting regional rural economies and their craft industries, rather than sacrificing them to large-scale industry, and in 1994 Norway voted to remain outside the European Community, in order to protect them from transnational capital. Interestingly, the leader of Norway's 'No to EU' campaign was computer pioneer Kristen Nygaard, co-inventor of the first 'object-oriented' computer language, SIMULA, which was originally designed precisely to support this kind of people-based economic planning.[26]

BEAUTY AND LOWER IMPACT, FROM THE BOTTOM UP

What would our environments be like if, instead of being used to eliminate skilled work, automation were used to support and expand it (as advocated by Nygaard, and in Britain by Mike Cooley, Howard Rosenbrock and other proponents of the 'socially useful production' movement – mentioned in Chapter 7)?

Some artists have shown what's possible (for example, the British artist Grayson Perry's richly detailed ceramics and tapestries – which take state-of-the-art, computerized looms to the limit of their capabilities, and draw big audiences, including people who don't normally go to art galleries[27]). In the 1970s, David Pye developed his own, much simpler techniques for putting machinery at the service of skill, including what he called a 'fluting engine' and other devices for taking wood-carving into realms of bewildering virtuosity. He believed machinery should be used to extend the 'workmanship of risk' in the way that good sports equipment allows climbers, for example, to tackle bigger challenges, without in any way eliminating the danger of failure. His fluted and turned wooden bowls and intricate, precise wooden boxes give a taste of the kind of world that is possible when machinery is an extension of skill rather than of management.[28]

The fact that the purpose of automation, and especially computerization, under capitalism has been to eliminate autonomous labor explains how we lost so much diversity – and how we succumbed to the idea that you couldn't have diversity (nice places to live and nice things around you) if you also wanted automation.

The architect Mark Jarzombek begins a book surveying the beautiful, efficient, low-impact architectures of 'first societies' (broadly, societies that are not based on agriculture) with the observation that 'organization of space is an integral aspect of human society, as fundamental as language and fire'.[29] The computer is a wonderful tool for exploring that language.

Christopher Alexander's 1977 book, A *Pattern Language*, applied the 'language' approach to architecture.[30] He drew from, and inspired, other disciplines as well, especially computer systems design. The book originated at Berkeley, California, amid the ferment of activity around computers, the new sciences of cybernetics and General Systems Theory[31] as well as Noam Chomsky's idea of 'generative grammars',[32] which says that the infinite richness of human language is built, bottom up, from finite numbers of words, assembled according to a compact set of innate rules that all humans share, whatever their culture.

Alexander and his collaborators showed that buildings can, and argued that they must, be created in the same way – starting from their smallest elements and the rules that govern them. Alexander's 'grammar' is described as a set of 253 'patterns', each of which:

> describes a problem that occurs over and over again in our environment, and then describes the core solution to that problem, in such a way that you can use the solution a million times over, without ever doing it the same way twice.[33]

The first of these is called 'Independent Regions'; others are named 'Degrees of Publicness', 'Old People Everywhere' and, getting down into the detail, 'Public Outdoor Room', 'Short Passages' and 'The Fire'.[34] Alexander argues that most traditional buildings and communities were built in more or less unconscious obedience to those rules, which is why older ones are so often preferred to modern ones (and his next book was called *The Timeless Way of Building* – 1979). Alexander maintained that:

> People should design for themselves their own houses, streets and communities. This idea... comes simply from the observation that most of the wonderful places of the world were not made by architects but by the people.[35]

The 'grammatical' approach can be applied as we get into the fine detail of our environments. The best-loved ones have a 'vocabulary' of constructional elements: particular kinds of doorways and porches, windows and window fixtures, bricks and blocks, particular combinations of materials and finishes, the ratios between one element and another, and so on, which can then be combined freely, producing the kind of pleasing variety and subjective feelings one experiences in an old town center, where a fairly limited palette of timber and brick types and finishes, ironwork, windows, and so on, has yielded an enormous variety of shapes and sizes of buildings, with no two exactly the same.

These environments were not and cannot be created by decree. They are typically the results of social processes of negotiation and communal discussion, and draw on conventions and routines that have been worked out in a similar way. This isn't necessarily always a harmonious process, but it never produces disasters on the scale that become possible when design is done top-down. And they imply approximate equality or at least the absence of unchallenged, self-confident dominance.

Computers could easily have aided a power-shift in that direction from the late 1970s onwards, had we been better at recognizing the opportunity and the threats to it. Computers can help individuals and communities surround themselves with the kind of richness otherwise only found in stately homes and so-called 'primitive' societies. This can happen when inequality retreats.

By the mid-1980s, just as the world was being reclaimed by the elites, cheap personal computers were making it possible for people to design and create dwellings for themselves that met their own needs better and fitted better into their surroundings than any corporate offering. For example, simple architectural design software existed for at least one of the first home computers (the BBC B – or 'Beeb') that did 'ray-tracing'. This allowed you to work out sight-lines for a building (so as not to impinge upon neighbors' outlooks), and the fall of sunlight and shadow in different spots at different times of the day and year. These features are now routine in the much more powerful Computer-Aided Design (CAD) packages used by large architecture and building firms – but for speedy, profitable construction that complies, where necessary, with government and local regulations, rather than making things nice for the neighbors and the plants.

A Beeb-owning friend of mine owned a small, sloping, irregular bomb-site and wanted to build a workshop on it. He used his computer to create pictures of it (colored in by hand) for each of his dozen or so neighbors, showing what it might look like from their own windows. Most people liked it; some had objections which he addressed, and then printed out revised drawings and took them around again, door to door. After a couple of weeks of this, everyone was happy. It took a lot of care and time, but nothing like the amount of care and time people put into their homes anyway during their lives, trying to make them look nice, as we say, but oblivious of what the thing looks like from the outside.

The main constraints on 'making a home look nice' were usually laid down long ago with a 4H pencil and ruler by someone in a legal organization, disregarding any undulations in the land, wildlife, customary uses, trees or small streams that get in the way. Most of us are stuck with our rectangular plots as if God had ordained them. However much money we spend on them, our aspirations can never go further than 'my pink half of the drainpipe'.[36]

If the kind of computing power the big firms now enjoy were deployed instead on behalf of individuals, households and groups of neighbors, townscapes could become as magical as the ones people pay large amounts of money to bask in for a week or two every year in places as varied as the Greek island of Mykonos or the Italian city of Florence.

Each dwelling built in this way would inevitably be different from its neighbors, yet complementary to them, in the way traditional buildings are all different and for the same reasons. They would fit in with their neighbors, and with the terrain, like pieces of a puzzle, enhancing its features, however modest they might be (instead of obliterating them, even when they are fairly substantial, as happens in current building practice thanks to the relative cheapness of earth-moving equipment and fossil fuels, compared to human labor).

This would be an obvious role for the 'plan factories' envisaged by Cornelius Castoriadis (mentioned in Chapter 12) and their scope need not (and should not) stop at buildings, but could also include transport, provisioning, maintenance and employment, leisure, childcare, healthcare... everything that comes within anybody's range of awareness. The calculations are complex, but the multi-party, multi-dimensional negotiations they involve are exactly the kind of challenge that gave rise to the invention of linear programming (described in Chapter 12) in the 1930s and 1940s, where large numbers of constraints had to be balanced against each other in turn in the search for a few optimal solutions. Personal computers made it possible for a single household within its own, small neighborhood to do the same kind of number-crunching calculations that could once only be done for massive national projects.

This is exactly the kind of scenario people quite correctly call 'Utopian', although they don't necessarily mean that as a compliment. Yet it is entirely practicable – in fact, far more practical (if only on grounds of resource use and sustainability) than anything the dominant, top-down approach to building human environments can offer. It seems unrealistic, however, if your yardstick is what powerful people might allow rather than what the terrain and available resources will permit. The more one lives in the shadow of power, the more 'Utopian' any idea for change whatsoever is likely to seem.

AN 'UNEXPLORED TERRITORY' AT YOUR FINGERTIPS

The small details and surface textures of a building may affect those who use it more than its overall design (to which one pays little conscious attention after the first visit or two): its doors, doorsteps and doorknobs, walls, corridors, windows, steps... places where you sit, stand, cook, read, find privacy, meet people... and, at the tactile level, the materials they are made of and how they have been finished. This fits with what is now known about our sense of touch: all of our other external senses are derived from it. A physiologist, Ashley Montagu, who wrote a book about the sense of touch in 1971 calls it 'a new dimension, a new discovery, and unexplored territory holding much promise of secrets yet to be revealed'.[37]

It also fits with what David Pye had said only a few years earlier about 'the extreme paucity of names for surface qualities [which] has quite probably had the effect of preventing any general understanding that they exist as a complete domain of aesthetic experience, a third estate in its own right, standing independently of form and color.' These qualities are largely the result of craft rather than design – yet 'In the last 20 years there has been an enormous intensification of interest in Design... But there has been no corresponding interest in workmanship'.[38] Were inequality to recede, such skilled craft would flow back into our environments in interesting and wonderful ways. (To start with, imagine the increased need for skilled repair and assembly work, when the outlawing of workplace exploitation has made sweatshop-produced, throwaway products impossible to manufacture.)

For Pye, the term 'workmanship' covered not only human craft, but the work of nature that one recognizes in naturally formed materials such as sea pebbles, marble and the grain of wood. It explains the almost universal preference for materials that have been longer in the making: for close-grained hardwoods, for example, in preference to loose-grained pine. Pye's world did not overlap with those of Benoit Mandelbröt (who discovered the detail-within-detail forms known as fractals – see Chapter 7) or Christopher Alexander, but his sensibilities did. He might have been talking about fractals when he wrote that craft carries design down to the limits of awareness and below, and speculated that:

> *The downward extension of design to the minutest scale of workmanship is governed by the same law which determines the appearance of a distant mountain or gigantic building, or... that the elements on the threshold of recognition are important at every range.*[39]

This finer level of detail is not always perceived consciously yet it is very important for a person's sense of well-being or even for health. Researchers in the new field of environmental psychology have found that people are attracted to and reassured by fractal shapes in the same way they are by natural ones – and not by the kinds of shapes one is surrounded by in a typical urban environment, which lack the important 'detail within detail within detail' quality.[40] Dementia sufferers are happier and less dependent in such environments, as are children with attention disorders.[41]

Work by a Canadian ecological economist, Jing Chen, links this to the 'universal law': the entropy law described in Chapter 9. He theorizes that the human mind 'being a product of natural selection, calculates the entropy level [of what it sees] and sends out signals of pleasure for accumulating and displaying low entropy, and signals of pain for dissipation of low entropy.'[42] So it is not surprising to find that people function better mentally, and recover from illnesses more rapidly, when they have access to natural rather than artificial environments, and in built environments made from natural materials.

Design, it could be said, has been adopted as a substitute for craft and the tactile qualities it creates. As practised under capitalism, and especially computer-assisted capitalism, it is quintessentially heteronomous and bland: design from afar, by others. Almost everything in a modern working-class home is produced that way, including the home itself: designed on a computer in some property-development company's headquarters with minimal regard for its ultimate setting or the people who will live in it.

SHRINKING ROADS, EXPANDING DIVERSITY

If we are in an anthropocene age, then today's road networks are its characteristic geological formations: visible, ubiquitous evidence of what happens when a society allows the course of its economic development to be decided by positional competition between unequal players. There is a self-creating drive to over-capacity in areas where the competition is intense, and depletion everywhere else.

Paradoxically, a more equal world would need far fewer roads, yet offer much greater freedom of movement. In history, the main limitation on human movement was never that the highways weren't big enough, but the laws imposed by dominant groups on subordinate ones to prevent them moving around. This was so in ancient empires, in feudal Europe, and today, when multi-lane highways proliferate at the same time as frontier walls and high-tech border-protection systems. Roads and travel restrictions have often arrived simultaneously. In his book *Imagined Communities*, Benedict Anderson pointed out that the 'opening up' of a country to colonization always went hand in hand with strict laws restricting freedom of movement. Many developing countries' first taste of modernity has been a metalled road, for the benefit of mining or logging companies, or for the military. Europe's own canal, rail and road systems were introduced to increase the profitability of production and distribution.

As explained in Chapter 6, positional competition demands infrastructure that can meet peak demands – and therefore stands largely idle the rest of the time. The new spare capacity then creates new opportunities for competitive concentration of offices, depots and so on, setting in train new congestion problems, requiring yet more infrastructure... and increasing the amount of travel the rest of us have to do, the amount of transportation needed, and the diversion of creative energy required, simply to maintain day-to-day existence.

When the veil of illusion that all this is necessary is lifted, as it is sometimes by accident, the reality can seem unbelievable. Recall the extreme redundancy of resources exposed by the Cybersyn project in Chile in 1972 (see Chapter 13) when it turned out that only a fraction of the country's trucks were needed to keep essential goods flowing as normal.

The Chilean example fits with what we know about other societies that enjoy greater equality. People have to commute less in countries that are more equal. Materials are moved around a lot less. As mentioned in Chapter 4, people also enjoy greater social mobility – they can change careers and jobs more easily. The 'faster is slower' effect goes into reverse. And because fewer, smaller fortunes depend on competitive speed there is more scope for people to explore varied technologies.

Cycleways are tiny indicators of the almost unimaginable

possibilities of a world with radically fewer highways, where people are nonetheless free and able to travel wherever they wish, when they wish. Cycleways are a characteristic of cities where power is already more evenly balanced between different social groups, along with high-quality public transport, which is sometimes provided free. The leading examples are all in more egalitarian places, such as Gröningen in the Netherlands, Freiburg and Karlsruhe in Germany, and the Canadian city of Vancouver – which, in addition to being more egalitarian than similar-sized US cities, is the only major city in North America that does not have a multi-lane highway running into its central district.[43]

Bicycles, and even horse-drawn vehicles, already turn out to be rather faster than cars, when due account is taken of the time a person must spend earning sufficient money to own one. In 1977, a member of the Adret collective calculated that:

> When you look at the hours a car can save you and the hours you spend paying for it, you start yearning for the days of cycling and walking. A worker who owns a car has to dedicate, each year, for its purchase, upkeep, repairs and insurance, at least 375 hours or about two months of work.[44]

In 2006 Conrad Schmidt, of Canada's *Work Less Party*, calculated that working to pay for his car consumed almost three times as much of his time as he spent actually driving it: 82 hours each month to travel 1,200 kilometers – an average speed of 14 kph, which is slower than cycling, as he explains in his book *Workers of the World, Relax!*[45] And James Boyce (whose work on the relationship between inequality and environmental impact was mentioned in Chapter 4) regularly asks his economics undergraduates to carry out this same calculation; they come to similar conclusions, and many of them become confirmed non-car owners.

Wind power is a major technology that was pushed to one side by the fossil-fuelled capitalist epoch, but it has never gone away or stopped developing.

Sailing ships were not eclipsed for straightforward reasons. The fact that steam vessels had to carry their own fuel was a major handicap. However, the steel, coal and then oil interests that would profit from steam were becoming powerful enough to shape the outcome. The Panama Canal, opened in 1914 after expensive political and military interventions had made the project possible, gave steam an advantage in the trade between the Atlantic and Pacific. Until that point, sail still carried a sizeable fraction of international cargoes[46] and its technology continued to evolve, with ever-bigger, faster vessels, culminating in huge, fast, steel-hulled barques with advanced rigs and power assistance for raising and trimming sail – such as the Hamburg-built five-masted barque *Preussen*, launched in 1902.[47] Sailing-vessel numbers fell during the two World Wars – they were more vulnerable to attack and did not fit the convoy system. But even so, tall ships returned to serious work in the Baltic after the wars were over, and a Hamburg-built four-master, the *Pamir*, built in 1904, was put to profitable transatlantic work by its owner and its enthusiastic crew until 1957, when it was overwhelmed – not by market forces but by a hurricane off the Azores.

Enormous possibilities are offered by new materials and computerized, servo-assisted rigs, and people are constantly looking for ways to bring them into service. For example, the Dutch yacht builder Dykstra was 'planning an entire armada' of large four-masted 'Ecoliners' in 2012[48] – although these seemed unlikely to offer a particularly radical departure from the present norm while they have to fit in with a maritime freight system based on containerization (and as of March 2016 the project was still only 'ready to leave the drawing board').

Aircraft have become synonymous with environmental destruction, but how much of their destructiveness comes from flight itself, and how much from the capitalist way of flying – governed as it is by maximizing speeds and payloads? Birds and insects can stay aloft all day, without worrying whether they can afford to or not, adding to the diversity of their environment while they are about it.

Modern passenger jets are claimed to be very efficient on the basis of how much fuel it takes to move one passenger one mile or kilometer. The giant Airbus A380, for example, claims 78 passenger-miles per US gallon, but calculations like these are based on the assumption that several hundred people are to be carried simultaneously, very fast, over distances of several thousand miles, and does not include any of the costs of enlarging airports, runways and other infrastructure to cope. In terms of shifting a given tonnage of aircraft a given distance, the latest jets are only slightly more efficient than piston-engined airliners were in the 1950s.[49]

If speed became radically less important, as it would do in a world without entrenched positional competition, that would alter all the bases of calculation. As the materials scientist JE Gordon pointed out (Chapter 8), some kinds of aircraft can be more durable than cars, and it is possible that smaller, lighter, slower aircraft might be more fuel efficient.[50] Airships would become viable, and these can have negligible environmental impact, as well as being a very different kind of experience.[51] These were the comments of BBC correspondent Anthony Smith, who took a flight on a prototype in 2007:

> We just cruised for 40 minutes, but could open the windows, speak without effort, enjoy watching the world go by 1,000 ft (300m) below, and tell ourselves what it must have been like when far bigger airships were having their heyday. Such as the Graf Zeppelin which went around the world in 1929 in four hops, starting from the US, touching down in Germany, then in Japan, and then in California.
>
> What a flight, with meals in the dining room, cabins to sleep in, and our beautiful planet not six miles down and invisible but usually a mere 1,500 ft (450m) below.
>
> Think of all such trips. Perhaps down to Rio in one hop, dancing if you felt like it, walking about, and not just to a doll's-house loo.[52]

Paradoxically, slower modes of transport might deliver a greater diversity of goods. For example, the practice of flying fruit and vegetables all over the world has gone hand in hand with a radical reduction in choice: a global system favoring varieties that can survive long periods under refrigeration without unsightly blemishes, and can be grown at different latitudes to take advantage of different growing seasons. At the beginning of the 20th century, 7,098 varieties of apples were known in the US.[53] Fifteen varieties now account for 90 per cent of all apples sold in the US, and the biggest sellers are varieties recently developed (and even patented) in New Zealand for the new, globalized fruit trade. One variety, Royal Gala, ranks second in the US and accounts for 20 per cent of all apples sold in the UK. Another variety bred for global trade, Braeburn, appears year round in British supermarkets, from New Zealand, France, Chile and sometimes from England as well.

Conversely, slow transport can have a surprisingly long reach. According to my mother (born 1919), English greengrocers in the 1930s had most of the same foreign fruit that supermarkets have now and some things that they don't: bunches of fresh mimosa would appear every February, having been brought overnight by train from southern France, which seems quite wonderful (and certainly seemed so to her). There were also many more kinds of more locally grown apples, plums, pears, even within my own lifetime. The labor content of such food is higher, but so is the nutritional value.

PUTTING BABIES AND CHILDREN AT THE HEART OF THE ECONOMY

The challenge is to restore the concept of 'work' to democratic debate and control – and make sure people can afford to do the work they value, and which their families and communities need and want. This Utopian notion would become a lot easier to contemplate, if the positional pressures that make life difficult and expensive were eased.

Any calculation of how much work would be necessary in an ideal society is bound to be flawed if it is based on industrial productivity, and the kind of work that is done in things called 'jobs', for wages. It ignores at least half of the world's work – the unpaid work that women usually end up doing. Capitalism likes to ignore all this, and assumes that its most valuable assets appear, ready for action, out of thin air: the able-bodied, clear-thinking, adaptable people it needs in order to function. This magical army is the product of 'reproductive labor': not just the business of making babies, but also bringing them up, clothing and feeding them, keeping roofs over their heads, providing warmth, food and pleasure, and so on. In the 1970s some radical socialist groups tentatively proposed placing such activities at the center of economic life. This is from a book by a group of trades councils in northern England:

> 'How would we organize the economy in order best to care for and support our children?' Not a question asked very much, if at all, in conventional economic policymaking... It implies a reversal of all the most central economic relationships which make up the taken-for-granted framework of policymaking.[54]

Ursula Huws has pointed out that this unpaid, reproductive workload increases when we allow 'work' to be defined as paid employment. So-called 'labor-saving devices' created almost as much work as they allegedly saved during the 20th century, and it is still mainly women who are doing it. She writes:

> Each housewife, isolated in her own home, duplicates the work of every other housewife, and requires her own individual washing machine, refrigerator, stove, vacuum cleaner, and all the other items that make up a well-equipped home, from lemon squeezers to deep-fat fryers, many of which are probably out of use 95 per cent of the time. There is thus no economy of scale, which is often the main saving that automation brings. Getting out the food processor, assembling the bits, dismantling it, washing it up, and putting it away again takes as much time whether one is cooking for 2 or 20, and the same applies to hundreds of other operations that all women carry out separately.[55]

In 2005, US women were spending 29.3 hours per week on average on 'home production' and men 16.8 hours – a total of 46.1 – not enormously different from the total for 1900 (50.7 hours).[56]

Huws notes a tendency among some leftwing writers to ignore all that, and see signs of a revolutionary future in the kinds of creative work ('cultural labor') that have blossomed among computer users: programming of various kinds, and now writing blogs, exchanging messages and organizing one's life via social media. But the housework still has to be done. She writes:

> A vision of the future that filters reproductive labor out of view runs the risk of failing to predict the next big wave of commodification. And a Utopia that focuses only on those activities that currently take place visibly within the market runs the risk of leaving the gender division of labor intact and disregarded. While Adam blogs, we must ask, who is cleaning the toilet?[57]

We need a real revolution in the design of *all* work. Radical though that sounds, it can start in ways that don't look very glamorous (like the co-housing project mentioned in Chapter 4), and ones that lurk on the fringes of society. Gypsies and travellers live wherever they can and their lives are hard, but they know about autonomy, they know their own and each other's skills and strengths, and they know where their livelihoods come from. This has made them a tough nut to crack for the authorities that want to get rid of them.

SHARED WORK: UTOPIA'S POWERHOUSES

Work that's shared feels radically different from work done in isolation. Stephen Marglin (in *The Dismal Science*, his damning analysis of his own discipline, economics) contrasts the harsh, isolated lives of women in the Texas hill country in the 1920s (described in Robert Caro's biography of President Lyndon Johnson, who grew up in that area) with the lyrical account given by writer Sue Bender of performing identical tasks in the Amish community where she went to live in the 1980s.[58] Among the Amish, community takes the highest priority. Hard and unappealing tasks are done communally and sociably if possible. They reject technologies that would subvert the communal life but embrace ones that support it.

Unfortunately, a great many technologies produced in the

outside, capitalist world seem specifically designed to eliminate communal activity. It would be commercial suicide for a capitalist firm to produce a washing machine that a number of people can share, when it could sell each person an entire laundry of their own. The same logic forces us all to buy our own computers instead of sharing them, even though they are built on technologies that were perfected for that very purpose (via time-sharing, as described in Chapter 11).

Eric Brende (a computer scientist who has lived in an Amish community) found that computers can be perfectly acceptable among the Amish because they are so useful for doing the farm accounts. But internet access in the home is generally not accepted, because of the ease with which it can draw a user into a different and possibly damaging world, separating them from their community.[59] Brende, interviewed in 2004, touches on an important difference between 'friendship' as understood in modern societies and as practised in, say, social media (between people you like rather than dislike) and the kind of social relationship, and communication, that comes out of doing necessary things together:

> People get together not just because they like each other, but because they need each other. There's a strong incentive not to sweat the small stuff.
>
> There's another whole layer of more subtle dynamics at work. When you are working with your hands, or whatever limbs, out in the field, pretty soon that work becomes self-automating. It thereby frees up the mind for conversation. Meanwhile, the labor serves as a kind of musical undercurrent that gives a certain depth to the experience. It's like the difference between hearing a choir singing in unison, and one singing in harmonies, with basses at the bottom.[60]

This bond between people that happens 'Not just because they like each other' is a constant and politically vital feature of solidaristic experiences, as we will see below.

But single-community solutions have their limits. All individual or local attempts to lead autonomous lifestyles are forced to exist in opposition to the wider world, unable to use most of its resources because they have been configured to undermine autonomy. But one can begin, perhaps, to imagine how different that wider world, and theirs, would be, if those resources were deployed to support autonomy instead.

COMMUNITY IS STRONGER THAN WE THINK: 'DISASTER UTOPIAS'

In A *Paradise Built in Hell* Rebecca Solnit describes a number of episodes when whole societies were briefly transformed by spontaneous, self-organized mutual help networks that sprang up from communities, in the immediate aftermath of disasters: explosions, earthquakes, hurricanes and the like.[61] She calls them 'disaster Utopias'. In all the disasters she researched (natural ones and human-made) the common factor was the sudden absence of normal, hierarchical organization. Autonomy immediately replaced heteronomy and proved much more effective. Accountability vanished and was replaced by trust, even in complete strangers, and it almost always proved to be more than justified.

She interviewed a young man, Tobin James Mueller, who set up a free coffee stall in Union Square, Manhattan, the day after the 9/11 terrorist attacks in 2001. His only previous experience had been organizing rave parties but his initiative somehow grew within a couple of days, into a highly efficient, 200-strong operation locating, collecting and delivering materials to firefighters and rescue workers, and organizing temporary housing. Mueller had made it a policy never to turn away volunteers (which official relief organizations did):

> A hopeful would-be volunteer comes up to me and asks if there is anything she can do. I give her a job and that's the last direction I need to give. Each volunteer becomes a self-motivated never-say-die powerhouse... they find 100 other more jobs to do... it's so much fun to participate in I forget to sleep... it's difficult to bring

*oneself to go back home... My one rule: I never say 'no'. That's one of the reasons it becomes a Utopia.*⁶²

Solnit found this to be a constantly recurring theme. It revolutionized people's lives, and some of those revolutions have continued. For example, the factory occupation movement that started in Argentina after the country's financial system collapsed in 2001-2 has matured and spread. When *New Internationalist* magazine's Vanessa Baird visited in 2013, she found 'a growing legacy of... "everyday revolutions". Of people interacting with each other, on an equal footing and with respect, to meet their needs, improve their lives and create a measure of social justice.'⁶³ This movement had spread far beyond Argentina, especially after the 2007/8 financial crisis. In February 2014 there were enough worker-controlled factories in France and Italy to justify an international meeting of the movement, which was held at a worker-controlled herb-processing factory in Marseille.⁶⁴

After every one of the disasters that Rebecca Solnit studied, the authorities and media predicted an outbreak of looting and 'bestial behavior'; what evolved instead was a riot of self-organized mutual aid, which was far more effective than anything the authorities could organize. The pattern was so pronounced as to call in question what we mean by 'normality': who is protecting whom against what?

Women found themselves doing 'men's' jobs, like firefighting, construction and organizing, while men lost their inhibitions about doing 'women's' tasks like feeding people, childminding and helping the injured.

The mutual-help networks that sprang up in the devastation of New Orleans in 2005, after the San Francisco earthquake of 1906, and many other disasters, had an almost festive quality, making the experiences perversely positive for many of those involved. All were suddenly equal. Nobody doubted anyone else's good intentions. Money became irrelevant. There was real work to be done – far more important and real than their day-to-day work had been – and everyone piled in to help with no bidding and tremendous efficiency. People found a sense of purpose they had craved all their lives and, having found it, were disinclined to accept anything less. After the San Francisco earthquake, wealthy people could not find servants. Lots of people were unemployed, but they didn't want the jobs.

During blackouts in Manhattan in August 2003, people turned out to help neighbors they'd never even spoken to before. Along with the air-conditioning, the street lighting went off, so people could see the stars for the first time, suggesting to Solnit that:

> *You can think of the current social order as something akin to this artificial light: another kind of power that fails in disaster. In its place appears a reversion to improvised, collaborative, co-operative, and local society… [The suddenly visible stars are remote] But the constellations of solidarity, altruism and improvization are within most of us and reappear at these times. People know what to do in a disaster.*[65]

There was nothing mystical about this eruption of trust and good humor. A fundamental biological system was at work here that capitalism systematically suppresses: the faculty that Heidi Ravven called 'the self beyond itself', which socialists call 'solidarity' and which Peter Kropotkin, the founder of modern anarchism, called 'mutual aid'. Kropotkin argued in 1902 that this is a major emotional mechanism found throughout the living world and, in fact, 'a factor of evolution' (the subtitle of his book *Mutual Aid*, written as a riposte to the idea that evolution proceeds by ruthless competition, advanced by Thomas Huxley in *Evolution and Ethics*, in 1893). In his introduction to *Mutual Aid*, Kropotkin is at pains to demolish what he considers a false opposition that has dominated attempts to understand human nature: it is either aggression or 'love that makes the world go round'. Kropotkin writes that:

> *to reduce animal sociability to love and sympathy means to reduce its generality and its importance, just as human ethics based upon love and personal sympathy only have contributed to narrow the comprehension of the moral feeling as a whole. It is not love to my*

> neighbor – whom I often do not know at all – which induces me to seize a pail of water and to rush towards his house when I see it on fire; it is a far wider, even though more vague feeling or instinct of human solidarity and sociability which moves me. So it is also with animals.[66]

A US neuroscientist, Donald Pfaff, referred to by Heidi Ravven, believes the response is the same one that creates the bond between mother and baby: a blurring between one's sense of oneself and of another person (the sense of empathy) mediated by identifiable interactions between neurotransmitter chemicals, oxytocin in particular, and specific networks of brain cells. He mentions a man called Wesley Aubrey who jumped down into the path of an oncoming New York subway train to rescue a total stranger who had fallen onto the track during an epileptic fit. Pfaff writes that:

> Mr Aubrey's brain must have instantly achieved an identity between his self-image and the image of the victim who fell in front of the subway train. This identification did not occur by some complex highly intellectual act – it came about by... blurring the distinction between the two images. In addition, Mr Aubrey was demonstrating the kind of prosocial caring feeling that (I hypothesize) normally develops from parental or familiar love.[67]

As everyone knows who has ever responded to another's need, or cried while watching a movie or even when reading some story of self-sacrifice, this 'loss of self' is a physiological phenomenon, and it happens faster than thought. It is a stark testament to the oppressive ideologies we have learned to accept, that we should need scientific approval to take seriously what John Donne wrote nearly four centuries ago, in the much-quoted paragraph from his Meditation 17 (written while he was recovering from typhus during the epidemic of 1624, and the funerals of other victims were taking place all around):

> No man is an Iland, intire of itselfe; every man is a peece of the

> Continent, a part of the maine... any mans death diminishes me, because I am involved in Mankinde, and therefore never send to know for whom the bell tolls; It tolls for thee.

Why has that little paragraph such emotional force? Everyone recognizes its truth, whether they know their Kropotkin or their neuroscience or not. Pfaff says we actually do recognize it in our guts. Pauline Jacobson, a survivor of the 1906 San Francisco earthquake, wrote:

> The individual, isolated self was dead. The social self was regnant. Never even when the four walls of one's own room in a new city shall close round us again shall we sense the old lonesomeness shutting us off from our neighbors.[68]

Rebecca Solnit says: 'The possibility of paradise hovers on the cusp of coming into being, so much so that it takes powerful forces to keep such a paradise at bay.'[69]

No wonder the Right invests so much energy, and takes such big risks, in destroying the power that comes from the solidarity that develops in workplaces.

THE RIGHT KNOWS THE POWER OF SOLIDARITY, EVEN IF THE LEFT DOESN'T

Even filthy, dangerous workplaces can be Utopias according to many who work in them, because of the solidarity that reigns there. One thinks of the dynamic role played in revolutions and struggles for justice by people who come from exactly those kinds of workplaces where people develop and take pride in their own powers: sailors, printers and miners in particular – tin miners in Bolivia, copper miners in Peru, coalminers everywhere. The Right is often better at recognizing their power than are politicians of the Left.

Britain's neoliberal prime minister Margaret Thatcher instinctively hated trade unions, whom she labelled 'the enemy within'. Defeating and neutralizing organized labor was central to her mission to reverse the post-War trend of reducing inequality, and defeating the coalminers was the central part of that task. Most of Britain's 200,000 coalminers lost their livelihoods soon after their defeat in a strike (1984/5) provoked deliberately by the government to destroy trade-union power.

Given the extreme discomfort and danger of coalmining work, it can seem extraordinary that people would fight so hard and suffer so much to preserve it – but perhaps it is no more extraordinary than the lengths dancers, musicians, writers and climbers will go to, to do what they have set their hearts on doing. It may seem odd to speak of coalminers in the same breath as ballet dancers and composers, but that may reflect more on the inequality of a society in which these different communities have so little mutual contact, than it does on the nature of their work per se.

Miners and their communities prized the solidarity and the equality that came from pit work. A Welsh miner my partner met during the strike spoke of his frustration at being restricted by ill health (caused by working underground) to surface work in a comfortable office. Underground, all were equal; management was left behind at the pit-head.

In 2009 I asked a Durham miner and union organizer whether he still missed working underground after all those years. As I recall, he said: 'I miss it terribly, every day'.[70] A few years later I asked another pit veteran what he thought about the man's remark. Could someone really feel that way about the pit?

'Oh absolutely!' he replied. 'There's no stress at the coal face. You're surrounded by terrible dangers, but there's no stress!' He described the camaraderie. It's impossible for anyone to try to pull rank because everyone is filthy and either naked or semi-naked (because it is hot underground). The humor is constant and hilarious: 'everyone takes the piss out of everyone else'. Rivalries and even animosities can be intense, but solidarity overrides them, instantly, when there's any external threat to any member of the group. 'All out! No two ways about it!'[71] John David Douglass's autobiography, *Geordies – wa mental*, overflows with the gleeful sense of strength that comes with a life of tackling real, tough issues with comrades – who one may not even like, but who are one's equals.[72]

A comrade is not necessarily the same thing as a friend.

Comradeship spells trouble for oppressors yet if they ever managed to get rid of it completely they would have killed the goose that lays their golden eggs. The merest scrap of Utopia can sustain the most downtrodden worker, more so even than food, as the following example shows: the art critic and social reformer John Ruskin found it in his *Daily Telegraph* in early 1864, and read it out in Manchester to a gathering of the great and good, to their annoyance. He published the lecture as *Sesame and Lilies*, and got the printers to print the passage in red (and it continued to be printed that way in later editions till at least 1927).

Michael Collins, aged 58, of Spitalfields, London, had died of starvation and overwork in the winter of 1863-4. He, his wife and son were 'translators' of boots: they collected worn-out ones from the rubbish tips, repaired them, and sold them to bootmakers' shops, which then sold them to people who couldn't afford new ones.

At the inquest, the coroner said to Mr Collins's son, Cornelius: 'It seems to me deplorable that you did not go into the workhouse.' Cornelius replied:

> 'We wanted the comforts of our little home.' A juror asked what the comforts were, for he only saw a little straw in the corner of the room, the windows of which were broken. The witness began to cry, and said that they had a quilt and other little things... In summer, when the season was good, they sometimes made as much as 10 shillings profit in the week...
>
> A juror: 'You are dying of starvation yourself, and you ought to go into the house until the summer.' –
>
> Witness: 'If we went in we should die. When we come out in the summer we should be like people dropped from the sky. No one would know us, and we would not have even a room. I could work now if I had food, for my sight would get better.[73]

The younger Collins' passionate rejection of 'state benefit', such as it was, expressed something much less comfortable than the terms 'hard-working families' and 'strong work ethic' convey, which the authorities like to applaud.[74] These were the defiant words of a human being, determined to preserve his last, infinitesimal scrap of the comradeship, warmth and dignity a human being needs, or die in the attempt.

The inquest jury very likely also knew that the workhouse wouldn't get anywhere near such good value from him as he'd produce, starving, through his own desperate efforts. Down the centuries, London's workhouses never produced much of a profit, if any, from their inmates, no matter how strictly regulated the regime.[75] The Victorian workhouse functioned, and was intended to function, mainly as a threat, to keep people working and maintaining themselves at their own expense – or have their humanity confiscated.

A market economy wrings serious profit from the tiny Utopias that sustain free labor – even while denying that they play any part at all in the proceedings, or that they even exist.

UTOPIA: NOT A WISPY ASPIRATION BUT A TOUGH REALITY

The big lesson of Rebecca Solnit's work – and even more, that of Petr Kropotkin and other anarchist thinkers, not to mention modern neuroscience and social psychology – is that Utopia not only exists but that we know *where and how* it exists. It resides in the absence of domination, and in the autonomy and solidarity that become possible when domination is absent. Utopia emerges spontaneously, rapidly and fully formed in the way a plant emerges from a seed, the moment hierarchies of power are out of the way or people are resolved to be rid of them. Those hierarchies know this by instinct, and invest large amounts of their own and society's time and resources in the battle to suppress Utopia and deny its existence, while trying to find ways of creating private Utopias for themselves.

Elites are uncomfortably if dimly aware that the ideas and technologies that give them their power and wealth emerge from exactly the kinds of Utopias they dread. Fortunately for them, the

Left in general is shy of Utopia and hopes only for 'a bit less rape', as it were, so the capitalists can play with Utopia to their hearts' content in their managerial Prospero-islands and billion-dollar creative playpens, without too much risk.

But no wonder the Left is timid. To abolish hierarchy has come to seem such a tall order that we would rather try to change our natures than to confront it – and we do this every day, every time we pass a beggar in the street. But an apparently rock-solid consensus in favor of elitism can evaporate when an 'egalitarian turn' is in the air. People who might otherwise oppose egalitarian policies can cease doing so, or support them, if that is where they believe their interests lie.

The turn towards greater equality in Europe and the US during and after the Second World War came about, not thanks to a society-wide Damascene conversion to egalitarianism, but because people of all shades of opinion and background saw and felt that extreme wealth was no longer generally approved. This vague but pervasive sense of where society was heading provided a sort of 'feed-forward' signal that, as can be seen in retrospect, started to shift society into what could have been a far less damaging course of development.

When the tide of opinion seems to be flowing against privilege and poverty, unlikely people can show support for the oppressed. Elites are exquisitely sensitive to threats to their legitimacy, and can seek to recoup moral ground by isolating and punishing members of their own class who push their elitism too far. In his book *Cotters and Squatters*, Colin Ward tells how, during the acute housing shortage in post-War Britain, property owners who tried to evict squatters, and officials who condoned the evictions, were vilified by some traditionally conservative town councils and newspapers, and the policy was changed.[76]

But how to change the sense of a society's direction? Robert Axelrod, who has spent his career studying co-operative behavior, found that for a social norm to become established, it is not enough for society to disapprove of those who deviate from it; society must also and especially disapprove of those who fail to show their disapproval of the offending behavior. This second-order norm-enforcement mechanism is called a 'metanorm'.[77]

In today's unequal world, metanorms tend to be used oppressively: they are what makes 'family honor' such a cruelly effective way of controlling the lives of young women. A father is condemned by the whole community for not keeping his womenfolk under proper control. Similarly, rightwing media can control the freedoms of trade unionists, migrants, single mothers, public employees and so on by vilifying those who support them, or who are thought likely to do so: social workers, school teachers, and especially any politicians who look as if they might be 'soft on' the deviance in question.

The 'zone of criminality' is made as wide and vague as possible, so that people must make strenuous efforts to avoid identification with it. The accusation of 'closet Marxist', 'bleeding-heart liberal' or 'fluffy do-gooder' can start a stampede for cover, leaving the 'benefit scroungers' and 'asylum cheats' isolated and defenseless. Liberal-minded politicians start competing to show how tough they are, and finally end up doing their opponents' dirty work for them.

But these tactics ought to work even better for the egalitarian Left than they do for the hierarchy-loving Right because that is where they evolved. The rich variety of 'counter-dominance' strategies observed in traditional societies by anthropologists such as Christopher Boehm (see Chapter 1)[78] for 'taking someone down a peg or two' with jokes, gossip and so on, are powerful mechanisms for maintaining their egalitarianism and health. No trace of an aura of approval or respect has a chance to develop around potential tyrants. A good metapolitical strategy might be to turn the spotlight of criticism not just onto its obvious targets, the 'fat cats' and super-rich, but also and especially onto those who curry favor with them, write their speeches, do their advertising and PR, or feature them as objects of admiration in the media.

In this way, a general realization can quite suddenly ripple right through society (as the Algerian activist and writer Frantz Fanon put it in 1960) that 'rich people are no longer respectable people'.[79] The subjective reality of living in a society can change totally, long before its physical structure does, producing the universal, galvanizing sense of having arrived in 'a paradise' or 'a new era', described by Gustave Courbet during the Paris Commune in 1871, and by George Orwell in Republican Barcelona in 1936. One moment oppression seems universally assumed to be the norm for now and evermore; next moment it is impossible to find anyone who was ever in favor of it. It is not that people are fickle and have changed their minds but that certain long-suppressed, unarticulated beliefs and feelings are suddenly free to come into the open.

But people will never have a sense that things might move in an egalitarian direction if nobody will declare publicly that equality plain and simple is what we need. And no new era will last beyond its dawn, if its people start to tolerate inequality again.

The likelihood of rightwing terror can never be dismissed, but solidarity is the best defense against it. That's why elites have always sought to stigmatize and destroy sources of solidarity. The struggles of mine workers and slum dwellers, not to mention soldiers, show that when people have solidarity they can endure things the mere thought of which would make a lone individual throw in the towel.

EQUALITY, TRUTH AND THE EXPERIENCE OF BEING BELIEVED

We can recognize Utopia by the feelings it evokes; in fact, without those feelings it is not Utopia. Equality, freedom from the fear of domination, is always the root of the matter.

One thing that stands out is how important truth-telling is in the lives of hunter-gathering and foraging peoples – and how important the converse, avoidance and denial of the truth, is to a modern society. Hunter-gatherers' candor can even be quite shocking. But theirs is a culture that cannot afford falsehood. Survival can depend on disclosure of every possible scrap and nuance of information. Hugh Brody believes that:

> *The apparent sturdiness of the hunter-gatherer personality, the virtual universality of self-confidence and equanimity, the absence of anxiety disorders and most depressive illness – these may well be the benefits of using words to tell the truth.*[80]

In class societies, the lower one's status, the less one can expect to be believed. The least powerful must work the hardest to prove that they are not lying. Credibility is an attribute of high status, traditionally males, especially white ones. Low-status people always have to prove that they are not lying, especially if they are ill or in need. Victims of sexual abuse and migrants are often the least likely to be believed, especially in the most unequal countries. The simple matter of being taken at your word – a basic assumption among 'primitive' people – is like a miracle. This was said by a woman who had been raped, quoted in a recent UK government report:

> You need someone to say 'I believe you'. That's the most important thing. Anything after that is great, but that's what screws your head: someone calling you a liar.[81]

The emotional impact of simply being believed at last, when you are used to being disbelieved and having everything you say challenged, is transformative and empowering. Another woman, describing her arrival 40 years ago in an overcrowded refuge for battered women, where her story was finally believed, said 'my life began in the refuge'.[82]

Credibility is fundamental to autonomy, and autonomy is anathema to a class society; it would not do for the lower orders to decide for themselves what they need, let alone 'help themselves' to food or housing, even when there is plenty of it lying around and going to waste. The poor and sick must submit their most intimate sufferings to tribunals, which decide whether those sufferings are real or not. Stigmatized minorities are non-credible by default. Subordinate workers have to justify taking time off work, or complete time sheets to prove their attendance, or even ask permission to go to the toilet. At the other end of the credibility spectrum, a superior might say 'I feel a bit rough this afternoon; I think I'll head off home now', and his colleagues and underlings will soothe and support him in that intention.

Inequality turns credibility into a sort of an invisible, official currency that touches the parts of our lives ordinary currency cannot reach. Instead of being something one simply has and can rely on, one finds that it lies in the hands of others.

In fact, you could say that 'credibility is like money' is more than analogy: it *is* money. As a number of writers have shown recently (for example, the anthropologist David Graeber[83] and the economist Ann Pettifor[84]), money, and especially credit (which is what nearly all modern money is, created by banks at the stroke of a pen on conditions that they define) is just a reification of the older notions of trust and reciprocity that humans always used among themselves. To find oneself beyond its reach is the very stuff of Utopia. To be believed, to be treated as if one's feelings, thoughts and experiences were as worthy of respect as anyone else's, to be given as much time as one needs: these are Utopia's basic ingredients. Compared to this, the physical conditions can be of minor importance.

Rightwingers are fond of warning about the utter mayhem that will ensue if social justice has its way, but the 'new heaven and new earth' can unfold with little apparent change to the physical situation (think of the transformation that can happen in a workplace, when the boss is away).

It might be a case, simply, of existing institutions starting to do what they already claim to do ('serving the community' and so on), rather than having to pretend all the time. One question worth considering is: how much of the existing structure of society comes from inequality, and how much of it comes from the Utopian impulses and dreams that keep the whole thing running?

There might even be a role for a marketing department in a firm run by equals, for equals: one that did what marketing people have always said they do: find out what people want, and then find ways of getting it to them.

WHO'S AFRAID OF PETER SAUNDERS?

Dare we demand equality? Given all the evidence, you would think that if we did we would be pushing on an open door. Few seem to object to the idea of *greater* equality. Even those who support the status quo don't explicitly demand *less* equality (although they back policies that lead in that direction). On the other hand, hardly anyone seems willing to suggest a level of inequality that would be 'about right'. The UK Green Party at least proposes a maximum wage of sorts (a 10:1 maximum ratio between top and bottom salaries in organizations) but why the factor of 10? Perhaps they think it is a realistic goal, one that powerful elites might not oppose as violently as they might 5:1 or 2:1, let alone equality plain and simple.

Instead of confronting the problem of excess wealth, liberal-minded political groups usually focus on relieving poverty, perhaps by enforcing and raising minimum wages. But raised minimum wages are easily negated when earnings and wealth at the top explode, driving up the price of housing and further augmenting the power of interests that are inimical to things that support general welfare, such as public transport, schooling and healthcare. As wealth gaps widen, the poor rely more on credit, which further enriches the already wealthy. Investment becomes increasingly focused on financial opportunities. The principle of inequality, unchallenged, becomes further entrenched.

The epidemiological evidence suggests that inequality is a bit like asbestos: it has no known 'safe level'. Why not ban it? Or why not at least discuss banning it, to draw out all the arguments pro and con? The writing on the wall suggests that the only safe level of inequality we should contemplate is zero. But why the silence? Are we at a historic moment, like one of those moments of stunned silence at the end of a stupendous performance, when nobody dares to be the first to clap?

Elitism's defenders have often argued that inequality is needed to spur innovation. The evidence gathered by this book contradicts that. The story of computers and high technology, in particular, tells us that tolerating inequality becomes downright dangerous as technology gets more powerful. The idea that big rewards (or even any material reward at all) are helpful in any productive sense, has no support from any of the relevant sciences and has been publicly demolished in thousands of academic studies and a slew of popular books (see Chapter 1).

Even some hard-liners are abandoning the claim that

inequality helps innovation. In 2009, the rightwing UK think tank Policy Exchange commissioned a veteran opponent of wealth redistribution, Emeritus Professor of Sociology Peter Saunders, to challenge the arguments made for reducing inequality (if not actually eliminating it) by Kate Pickett and Richard Wilkinson in their book *The Spirit Level*. He conceded that:

> No association between any of [the] indicators of economic vitality and the degree of equality or inequality of incomes in a country can be identified [and] a lot of defenders of free-market economics might also be wrong in arguing that radical income redistribution will necessarily choke off the spirit of enterprise and innovation in a country.[85]

Saunders made some attempt to show that inequality did not cause higher levels of morbidity, mortality and crime but Pickett and Wilkinson easily exposed his reasoning as highly selective and self-contradictory.[86]

They said nothing, however, about the stream of judiciously modulated abuse in which Saunders's objections came wrapped – as one might reasonably, in different circumstances, ignore someone's *faux pas*, either to avoid a scene or to avoid humiliating them. But this was not just a regrettable lapse by someone who knew no better. It was a sustained attack on Pickett and Wilkinson and those around them, whom he cast as 'leftwing intellectuals' with covert 'ideological' motives:

> **The Spirit Level** *is more than just an academic book. It is a manifesto. Its apparent 'scientific' backing for a core, traditional element of leftwing ideology is being used to spearhead a new political movement aimed at putting radical income redistribution back at the heart of the political agenda.*

Saunders asserted that because an 'agenda' lay behind their work, they were a discreditable source, part of a clandestine leftwing movement; anything they produced was bound to be tainted and it would be dangerous to base policy on it. Setting aside the question of whether Saunders might have had an 'agenda' himself, or represented any kind of political movement, the real force of his article was its innuendo (the inverted commas around 'scientific', the continual identification of the authors and all those around them as 'leftwing' and 'left-leaning'). Pickett and Wilkinson were not actually called 'reds under the bed' but when the factual discussion was stripped away, what you had left was classic rightwing intimidation in academic language, with footnotes.

THE 'APPARATUS OF JUSTIFICATION'

The French economist Thomas Piketty's major 2014 study of inequality, *Capital in the Twenty-First Century*, warns that global inequality is entering unknown territory with 'potentially terrifying' consequences.[87] In the US, income inequality is on course to set a new world record by 2030, with 60 per cent of all earnings going to the wealthiest 10 per cent.[88]

No society, says Piketty, has ever survived that kind of inequality without some kind of breakdown or revolution. Whether our present situation will prove sustainable or not depends to a large extent on what he calls 'the repressive apparatus' – which is indeed terrifying and unprecedented. It consists of militarized police forces, equipped with everything from tazers to drones and firearms of astonishing destructive power, backed by authorities who take an increasingly indulgent attitude to torture and imprisonment without trial. But also, before we ever get that far, we have to face rightwing media that have become expert at singling out potential enemies and useful scapegoats, hounding and humiliating them and those around them so effectively that nobody wants even to be seen anywhere near them, lest they get the same treatment.

But Piketty goes on:

> Whether such extreme inequality is or is not sustainable depends not only on the effectiveness of the repressive apparatus but also, and perhaps primarily, on the effectiveness of the apparatus of justification. If inequalities are seen as justified, say because they

> *seem to be a consequence of a choice by the rich to work harder or more efficiently than the poor, or because preventing the rich from earning more would inevitably harm the worst-off members of society, then it is perfectly possible for the concentration of income to set new historical records...*
> *I want to insist on this point: the key issue is the justification of inequalities rather than their magnitude as such.*[89]

I hope Piketty is correct to say that the apparatus of justification is the core problem, because (as I hope this book has shown) the very thing so many people see as capitalism's greatest justification, its claim to have given us a benign technological revolution, is a sham. In practice, huge media, industrial and financial interests do not want that claim undermined and will defend it tooth and nail with all the means at their disposal. But the facts are emerging, and bit by bit they are making their way into public consciousness.

Saunders's attack on *The Spirit Level*'s authors shows that the justificatory and repressive systems are not separate. Clear intimations of the scapegoating that precedes outright oppression are there in the 'measured' rightwing language. It is meant to undermine them, to prejudice readers against them, and to warn people off supporting them.

Verbal undermining is fundamental to the maintenance of inequality and injustice. Not challenging it or even drawing attention to it hands power to those who wish to preserve privilege and oppression. Pretending that one's attackers are fellow seekers after truth who share one's own values when it's clear that they regard them with contempt, betrays the entire constituency of the oppressed. It also hands oppressors and their helpers *carte blanche* to waste everyone's time with spurious or mendacious objections, which they can produce in endless quantities – demonstrated by the successful rearguard actions waged over decades by the tobacco industry against the evidence of its role in destroying the health of whole populations, and now by fossil-fuel interests against the evidence on climate change.

The linguistic scientist George Lakoff abandoned academia in the 1990s to throw the spotlight on how rightwingers win arguments by defining the terms and the language of the debate, and how the Left becomes complicit in its own defeat when it fails to challenge their basic tenets. As Lakoff has explained, rightwing values are known to be toxic (discipline, authority, punishment), yet rightwingers are completely open about them, shouting them from the rooftops, declaring that they are good and necessary, and defying all the evidence to the contrary. Facts are far less important in their world-view than values.

People of the Left ('liberals' in Lakoff's terminology) do the exact opposite. They tend to focus on the facts, as if this will get us out of dangerous, emotional territory, and soft-pedal on values for the same reason. Liberals then become complicit when they couch their own arguments in illiberal terms, on the grounds that these are now the terms of debate and it is impossible to do anything else in the current climate. For example, instead of plainly opposing workfare, they offer some ostensibly less cruel form of it, claiming that it will be a more effective way to coax more people into jobs – and justify their failure to call cruelty by its proper name by claiming that they are being realistic, because public opinion has shifted so far to the right that open opposition would be political suicide.

In the years after the Chilean putsch and the Thatcher/Reagan period, social democratic governments across the wealthy bloc of Western countries opted increasingly for this tactic for holding onto political power – from Bill Clinton in the US through Tony Blair in Britain to François Hollande in France. All ended up supporting tendencies most of their supporters found insupportable. All embraced the language of 'toughness' as applied to the poor and precarious, and became brave opponents of power only when the power concerned was that of trade unionists or of human rights groups.

Many people holding positions of responsibility for the state provision of healthcare tried to keep some control of the agenda by finding their own ways to enforce performance targets, pursue one-sided 'partnerships' with for-profit hospitals, and cut staff – and often sanctioned those who spoke out against what was happening. Many, perhaps most, campaigners for the rights of migrants and refugees believe the greatest problem is immigration controls themselves, which sooner or later will have to be scrapped. But many of them fear to say so openly and even turn angrily on those who do, saying that doing so could be 'counterproductive' because it would 'put people off' and 'put us beyond the political pale', or is even 'too advanced for the working class'.[90]

TELLING THE TRUTH

Inequality and the injustices and waste that go with it have gone from strength to strength since the landmark year of 1973, not despite these careful, 'realistic' tactics, but because of them, according to George Lakoff's analysis. To change anything we need to state our values openly. Unless we do so, nobody will be able to agree with them. We may find that the real 'silent majority' has been longing for clear statements of humane political principle, and will support them despite what the feared rightwing media may say. This has been demonstrated over and over again recently in the huge turnouts in elections wherever a genuinely radical alternative was being offered: in Greece, in Spain, and in Britain, where the landmark event was the Scottish referendum on independence in 2014. Many observers noted that the referendum had brought politics back to life and engaged the entire population because policies were being discussed that no Westminster party had dared to raise in public for some decades: free education at all levels; renationalization of public assets; an end to privatizations; a relaxation of anti-immigrant laws; and redistribution of land.[91]

So it's important to state what we want: equality, not just less inequality by some vague amount. Henceforth any inequality will need to be justified. This is not such an enormous conceptual leap from where we are now. Significantly, some of the most straightforward objections to excessive wealth have come from successful entrepreneurs. For example, the late Klaus Zapf, an energetic and successful entrepreneur who built up Germany's biggest removals firm yet lived in a small flat on a modest, worker's salary, is quoted as saying: 'I don't need money; it just makes us unequal'.[92]

There is an objection which points out that absolute equality is not achievable. That has never stopped societies outlawing other intractable injustices such as rape, murder, apartheid or even slavery. All the great battles against injustice, in which modern societies take such pride, were considered unwinnable until, suddenly, they were won. And some of society's most mundane underpinnings depend on equally 'unachievable' goals. Perfect verticality is a completely unachievable abstraction but it does not stop bricklayers continually checking that walls and lintels are as vertical and level as they can possibly be; we would never trust a bricklayer who did anything else. Nor does it stop us riding bicycles; and no ship's crew would throw themselves overboard in despair on learning that it is impossible to keep a seagoing vessel on a completely even keel.

No living organism or community of organisms has ever existed in a state of perfect and unchanging equilibrium until it was thoroughly dead. Life is always in 'a steady state of balanced tension'[93] – and that's mainly why we like it. However, life *does* approach permanent equilibrium at the bottom of a power hierarchy, which is why it is so often described as a 'living death'.

Let us agree not to restrict each other's aspirations in deference to what we feel we can get away with. Let's assume that the commitment to human equality that's written into the Universal Declaration of Human Rights means exactly what it says, and take it from there.

1   *A Viable Food Future, Part 2*, Utviklingsfondet (The Development Fund, Norway), Nov 2011, p 28, nin.tl/viablefuture , citing *The Ecological Wealth of Nations*, Global Footprint Network, 2010.
2   James Lovelock, *Gaia: A new look at life on Earth*, Oxford University Press, 1979.
3   *A Viable Food Future*, op cit, Part 2, pp 24-26.
4   Damian Carrington, 'Earth Has Lost Half of Its Wildlife in the Past 40 Years, Says WWF', *The Guardian*, nin.tl/wildlifeloss. Accessed 7 Dec 2014.
5   George Packer, 'Change the World: Silicon Valley transfers its slogans – and its money – to the realm of politics', *The New Yorker*, 27 May 2013.
6   For some powerful accounts of individual workers' and campesinos' experiences, see Colin Henfrey and Bernardo Sorj, *Chilean Voices: Activists describe their experiences of the Popular Unity Period*, Branch Line, 1977.

7   Timothy Garton Ash, 'It Always Lies Below', *The Guardian*, 8 Sep 2005. Quoted in Rebecca Solnit, *A Paradise Built in Hell,* Penguin US, 2010, p 241.
8   Jamie Peck, 'Neoliberal Hurricane: Who Framed New Orleans?' *Socialist Register* 43, no 43, 19 March 2009, nin.tl/neoliberalhurricane
9   Larry Elliott, 'Richest 62 people as wealthy as half of world's population, says Oxfam', *The Guardian*, nin.tl/richest62
10 George Monbiot, 'The British Thermopylae', 28 Aug 2014, nin.tl/Thermopylae
11 Conrad Schmidt, *Workers of the World, Relax: The Simple Economics of Less Work*, Work Less Party, 2006. Introductory note by Professor Christopher Shaw.
12 Stephen A Marglin, 'Premises for a New Economy' *Development* 56, no 2, June 2013, 149–154.
13 *A Viable Food Future*, op cit, Part 1, pp 27-29.
14 Shifa Mwesigye and Salena Tramel, 'Building a Peasant Revolution in Africa', La Via Campesina, 2 Oct 2013, nin.tl/peasantrevolution
15 *A Viable Food Future*, op cit, Part 1, pp 40-43.
16 FH King, *Farmers of Forty Centuries*, Mrs FH King, 1911, p 10.
17 T Schumilas, 'Alternative Food Networks with Chinese Characteristics', Doctoral Thesis, University of Waterloo, Ontario, 2014.
18 G Sturt, *The Wheelwright's Shop*, Cambridge University Press, 1930, pp 16-17.
19 M Greensted and M Batkin, *The Arts and Crafts Movement in the Cotswolds*, Alan Sutton Publishing, 1993, p 13.
20 Marshall Sahlins, 'The Original Affluent Society', in John M Gowdy, *Limited wants, unlimited means: a reader on hunter-gatherer economics and the environment*, Island Press, 1998, p 23.
21 HG Wells, *A Modern Utopia*, 1905.
22 Larry Elliott, 'Economics: Whatever Happened to Keynes' 15-Hour Working Week?' *The Guardian*, 1 Sep 2008, nin.tl/Keynes15hour
23 Adret Group, *Travailler deux heures par jour*, Éditions du Seuil, 1977, p 132 (author's translation). A summary in English can be found at nin.tl/work2hours
24 Wayne Roberts, *No Nonsense Guide to World Food*, New Internationalist, 2013, pp 102-104.
25 Robert Tressell, *The Ragged Trousered Philanthropists*, 1914.
26 Åke Sandberg, *Nordic Lights* SNS Förlag, 2013, nin.tl/Nordiclights Accessed 4 March 2014. See also Kristen Nygaard's address to the 1996 Information Research in Scandinavia (IRIS) conference: 'We are not against Europe. We are against Norwegian membership in the European Union', nin.tl/Nygaardlecture
27 'Craft in the information age', The British Museum blog, 15 Sep 2011, nin.tl/craftBMblog
28 Some examples of Pye's work are shown here: nin.tl/Pyework See also a description of Pye's approach by Simon Olding, University of the Creative Arts, June 2009, nin.tl/lookingbackwards
29 MM Jarzombek, *Architecture of First Societies*, Wiley, 2014, p. ix.
30 Christopher Alexander, Sara Ishikawa & Murray Silverstein, *A Pattern*

Language, Oxford University Press, 1977.
31 'Christopher Alexander', Wikipedia, 13 August 2014, nin.tl/1RaMjtG
32 Noam Chomsky's theory of 'generative grammars', published in 1958, argues that all human languages are generated by a brain structure, common to all human minds, that allows a finite number of rules to turn a finite number of words into an infinite variety of utterances. The Russian linguist Vladimir Propp had had a similar insight in the 1920s about the way folktales are created. Propp's book *The Morphology of the Folktale* (Leningrad, 1927) was first translated into English in 1958 – the very year Chomsky published his own seminal work. Neither, however, was aware of the other's work. Propp's work has been the inspiration behind many computer-based 'story generators' – which have numerous uses, for example, in computer games.
33 Alexander et al, op cit, p x.
34 For the full list of patterns, see nin.tl/patternlanguage
35 *A Pattern Language*, dustflap text, quoted by Wikipedia.
36 Vivian Stanshall's 'My pink half of the drainpipe' (1968) was released on the Bonzo Dog Doo-Dah Band's *The Doughnut in Granny's Greenhouse*, Liberty Records (UK):
My pink half of the drainpipe
I may paint it blue
My pink half of the drainpipe
Keeps me safe from you!
37 Ashley Montagu, *Touching: the human significance of the skin*, Columbia University Press, 1971, p 311.
38 David Pye, *The Nature and Art of Workmanship*, Studio Vista, 1964, p 99.
39 Pye, op cit, p 68
40 Caroline Hagerhill et al, 'Fractal dimension of landscape silhouette outlines as a predictor of landscape preference', *Journal of Environmental Psychology*, June 2004.
41 MG Berman, J Jonides & S Kaplan, 'The cognitive benefits of interacting with nature', *Psychological Science*, 19 (12), 2008, pp 1207-12, and sources referred to by them.
42 Jing Chen, *The Physical Foundation of Economics,* World Scientific Pub, 2005, p 11.
43 Conrad Schmidt, *Workers of the World, Relax: The Simple Economics of Less Work*. Vancouver: Work Less Party, 2006. p 117.
44 Adret Group, op cit.
45 Conrad Schmidt, op cit, p 117.
46 M Beenstock & A Vergottis, *Econometric Modelling of World Shipping*, Springer Science & Business Media, 1993.
47 Lance E Davis, Robert E Gallman & Karin Gleiter, *In Pursuit of Leviathan*, University of Chicago Press, 2007, p 262.
48 'Hybrid Container Ship Wind-Driven With "Automatic" Sails', TreeHugger, nin.tl/hybridcontainer Accessed 27 Sep 2014.
49 'Fuel Economy in Aircraft', Wikipedia, nin.tl/fueleconomyaircraft Accessed 11 Dec 2014.
50 BH Carson, 'Fuel Efficiency of Small Aircraft', in AIAA Aircraft Systems Meeting, Anaheim, California, 4-6 Aug 1980.

51 John Rennie, 'Does Global Warming Help the Case for Airships?' nin.tl/airshipscase Accessed 17 Dec 2014.
52 Anthony Smith, 'Flying from a Different Perspective', BBC, 13 Oct 2007, nin.tl/flyingBBC
53 *A Viable Food Future*, op cit, Part 1, p 26.
54 Coventry Trades Council, *State Intervention in Industry: A Workers' Inquiry,* Spokesman Books, 1982, p 162.
55 Ursula Huws, *The making of a cybertariat: virtual work in a real world*, Monthly Review Press, 2003, p 37.
56 Ursula Huws, 'When Adam Blogged', in *Gender and Creative Labour*, ed Ros Gill & Stephanie Taylor, Wiley, 2014.
57 Ibid.
58 Sue Bender, *Plain and Simple*, Harper, 1989.
59 Eric Brende, *Better Off*, HarperCollins, 2004.
60 John Zmirak. 'The Simple Life Redux: An Interview with Eric Brende', *Godspy*, 1 Nov 2004, nin.tl/Brendeinterview
61 Solnit, op cit.
62 Solnit, op cit, p 206.
63 Vanessa Baird, 'Argentina's challenge: Turning trouble into triumph', *New Internationalist* 463, June 2013.
64 'Report from the "Workers' Economy" International Meeting, January 31 and February 1, Occupied Factory of Fralib, Marseille', nin.tl/Marseillemeeting Accessed 9 April 2014.
65 Solnit, op cit, p 10
66 Petr Alekseevich Kropotkin, *Mutual aid, a factor of evolution*, Extending Horizons Books, 1955, p 6.
67 Quoted by Heidi M Ravven, *The self beyond itself an alternative history of ethics, the new brain sciences, and the myth of free will*, New Press, 2013, p 376 from Donald W Pfaff & Edward O Wilson, *The Neuroscience of Fair Play* Dana Press, 2007, pp 202-3.
68 Solnit, op cit, p 148
69 Solnit, op cit, p 7.
70 Conversation with David Douglass at The Anarchist Bookfair, London, 2009.
71 Conversation with Paul Winter at Ruskin College Oxford, 18 September 2014, at a meeting in support of Orgreave Truth and Justice Campaign, otjc.org.uk
72 DJ Douglass, *Geordies – wa mental,* Read'n'Noir, 2008.
73 John Ruskin, *Sesame and Lilies*, Dent, 1907, pp 35-37.
74 Jim Pickard, 'Tory Ministers Try to Flog the Phrase "hardworking" to Death', *Financial Times*, 30 Sep, 2013, nin.tl/FThardworking
75 LB Luu, *Immigrants and the Industries of London, 1500-1700*, Ashgate, 2005.
76 Colin Ward, *Cotters and Squatters: housing's hidden history,* Five Leaves, 2005.
77 RM Axelrod, *The Complexity of Cooperation,* Princeton University Press, 1997.
78 Christopher Boehm, *Hierarchy in the forest: the evolution of egalitarian behavior*, Harvard University Press, 2001.

79 Frantz Fanon, *The Wretched of the Earth*, Grove Press, 1965, p 154.
80 Hugh Brody, *The other side of Eden: hunters, farmers, and the shaping of the world*, North Point Press, 2001, p 195.
81 Taskforce on the Health Aspects of Violence Against Women and Children, *Responding to violence against women and children: the role of the NHS* [The Alberti Report] Available at: nin.tl/Taskforce, 2010.
82 *Woman's Hour*, BBC Radio 4, 15 Nov 2014, nin.tl/BBCNov2014 Accessed 18 Dec 2014.
83 David Graeber, *Debt: the first 5,000 years*, Melville House, 2011.
84 Ann Pettifor, *Just Money: how society can break the despotic power of finance*, Commonwealth Publishing, 2014. Also Mary Mellor, *The future of money from financial crisis to public resource*, Pluto, 2010.
85 Peter Saunders, *Beware False Prophets,* Policy Exchange, 2010.
86 The Equality Trust, 'The Authors Respond to Questions about *The Spirit Level*'s Analysis', July 2010, nin.tl/SpiritLevelresponse
87 Thomas Piketty, *Capital in the Twenty-First Century*, Harvard University Press, 2014, p 571.
88 Piketty, p 264
89 Piketty, p 264
90 This was the position of the British Socialist Workers' Party in the early 2000s, when it was building an alliance with the maverick ex-Labour MP George Galloway to form the short-lived Respect Party. An organization called 'Strangers into Citizens', which campaigned for a limited regularization of undocumented workers, also had heated arguments with Open Borders groups at around the same time. Yet both organizations agreed privately that immigration controls were in principle wrong and immoral, and should go.
91 Jonathan Freedland, 'Scotland Started a Glorious Revolution' *The Guardian*, 19 Sep 2014, nin.tl/Westminsteranoraks
92 'Eccentric German Millionaire Who Lived on Less than £300 a Month Dies', *Mail Online*, nin.tl/KlausZapf Accessed 13 Dec 2014.
93 William Gray, Frederick J Duhl & Nicholas D Rizzo, *General Systems Theory and Psychiatry,* Little, Brown, 1969, p 12.

Index

abandonment of technology 22
Ackoff, Russell 270
Adamatzky, Adam 232
Adorno, Theodor 50
Adret collective 310, 320
advertising 8, 171, 172, 214, 215-16, 221, 224, 264
agriculture 86
   crop diversity 322-3
   peasant 99, 103, 149, 307-9
   and population 98-9, 101-3, 149
aircraft 176, 321-2
airships 322
Ajax 224, 225, 226
Albania 77
Alexander, Christopher 313, 317
Alfani, Guido 56
Algeria 336
Algol 241
Allen, Robert 149
Allende, Salvador 263, 280, 281, 283, 285, 288, 293, 297, 299, 300
Altieri, Miguel 299
Alto 246
Amazon 214, 216, 224, 231
Amish community 325-6
analog computers 231-7
Anderson, Benedict 319
Andorra 95
Andreesen, Marc 221, 222-3
Anthropocene 97, 318
anthropometric data 84
Antonov, O 262
Apple 15, 17, 37, 42, 43, 57, 58, 185, 194, 214, 227, 246, 247, 248, 271
Arab empire 68
archeological data 24-5, 65, 84, 97-98

architecture *see* buildings design *and* Von Neumann architecture
Argentina 101, 328
ARM microprocessors 140
armed forces 85
ARPA 178
Arthur, Brian 19, 20, 23, 24, 154, 207
Ascherson, Neal 48
Ashby, W Ross 272, 273, 275
Ashby's Law *see* Requisite Variety, Law of
assembler 243
attenuation 275
Aubrey, Wesley 330
Australia 101, 139, 195, 199, 221, 310
Austria 95
automobile industry 32, 48, 163, 172, 172-5
autonomy 81-3, 94, 142, 164-7, 230, 254, 257, 281, 287, 290, 325, 327, 334, 338
aviation industry 32, 108, 176
Axelrod, Robert 335

Babbage, Charles 20, 38, 231, 234
Baird, Vanessa 328
banking 48, 156, 159, 163, 224, 228, 282, 285
Baran, Paul A 170, 171, 172, 187
Barbrook, Richard 272
Barker, David 87
BASIC 12
Bateson, Gregory 272
Battelle Memorial Institute 33
Beer, Simon 296
Beer, Stafford 266, 267, 268, 269, 272, 276, 286, 288, 289, 290, 292, 293, 294, 296, 298

Belgium 56, 61, 63, 95
Bender, Sue 325
Benner, Chris 141, 183-4
Berg, Raissa 86
Berger, Peter 37
Bernal, JD 267
Berners-Lee, Tim 210, 218, 219
Berreby, David 28
Berry, R Stephen 172, 173, 175
Bettencourt, Liliane 171
Bevan, RW 268
Bigelow, Julian 238
binary systems 231, 240
biodiversity 8, 99, 113, 303
Bion, Wilfred 49
birth defects 86, 87, 197
Black, Douglas 74, 81, 87
Black, Fischer 157, 158
Black, Michael 158
Black Committee report 74
Black Panther Party 270
Blackett, Patrick 267
Blair, Tony 344
Bloch, Marc 66, 265, 266
Boehm, Christopher 26, 28, 336
Boldrin, Michele 154
Bolivia 331
Boulton, Matthew 111
bow-wave effect 129-30
Bowlby, John 49
Boyce, James 96, 320
brain hemispheres 36-7
Brand, Stewart 185
branding 171
Braudel, Fernand 58, 60, 63, 69, 265
Brazil 101, 212, 308
Brende, Eric 326
Bretton Woods system 55, 282
Bricklin, Daniel 42, 147
Bridge, Gavin 114, 200, 201

Britain 8, 9, 31, 34, 51, 60, 67, 68, 73, 95, 101, 106, 114, 118, 124, 134, 147, 158, 159, 160, 163, 212, 215, 221, 231, 241, 264, 266-7, 283, 286, 290, 309, 344, 345
  defense 38, 39-40, 85, 162, 235, 238, 266-7
  education 137-8
  health 74-5, 81-2, 84-5, 87, 98, 143, 256
  housing 105-6, 137
  technology development (1830s) 20, 38; (1900s) 38; (1940s) 39-40; (1990s) 240
  trade unions 331-2
  wealth inequality 56, 283
  welfare state 48, 76, 87, 143
Broadhurst, SW 40
Brody, Hugh 26, 337
Brookings Institute 42
Brousentsov, NP 243
browsers 220, 221
buildings design 128, 167, 313-15, 316-18
Burawoy, Michael 30
Burch, Noël 107
Burma 60
Bush, Vannevar 218, 235, 236
Butler, Kiera 194

C language 240
CAD *see* Computer-Aided Design
Cafod 195
Cairncross, Frances 135
call centers 107, 141
Cambodia 148, 153
Campbell-Kelly, Martin 230
Canada 101, 108, 199, 212, 227, 298, 320
Canon 247
capital, redundancy of 293

Index

capitalism 24, 29-31
  and innovation 31-3, 35-7
  see also neoliberalism
Capitalist Calculation Debate 262
Carleton, Will 100
Carlson, Chester 33
Carnot, Sadi 203
Caro, Robert 325
cars 51, 76, 105, 112, 126, 130, 131, 132, 135, 138, 172-5, 320
  see also automobile industry
CASE tools 165
Castoriadis, Cornelius 166, 253, 256, 274, 315
Castro, Fidel 297
Cat 247
centralization
  of economic planning 251-62
  of organizations 119
  of resources 66-7
Centre for Unconventional Computing 232
CERN 218
Chicago school of economics 281
children 8, 49, 57, 74, 83, 84-5, 246, 284, 310, 324
Chile 257, 263
  Cybersyn project 275, 276, 286-92, 293, 295, 304, 319
  October strike 293-6
  overthrow of Allende government 281, 297-300
  Unidad Popular 283-6
China 29, 56, 60, 63, 68, 80-1, 87, 102-3, 108, 153, 195, 197, 212, 283, 308-9
  Foxconn factory 15, 58-9, 107
Chomsky, Noam 313
Christensen, Clayton M 185, 186, 187

Christiansen, Eric 62
Churchill, Winston 48, 267
Cisco 214
city states 61
civil service 8-9, 81-2
Claxton, Guy 35, 141
climate change 303, 343
climate crises 30, 97, 108
Clinton, Bill 344
Clissett, Phil 309
clock-driven systems 238-42
clothing industry 57, 59, 62, 107, 125, 152-3
cloud computing 55, 212-17, 247-8
Cockshott, Paul 263
code-breaking 39-40, 235, 238
Cohen, Daniel 34
Cohen, Nick 15
Colcernian, Ron 248
Cold War 42, 176, 177, 237, 272
Coleago 264
Collins, Cornelius 333-4
Collins, Michael 333
Colossus 39-40, 238
coltan 198-9
commodification
  and intellectual property 151-3
  microchips 178-82
  and positionality 149-50
  of work 107, 150, 161, 162-4
Commoner, Barry 299
Communism 51-2, 53, 256, 281
commuting 135, 160, 319
comparative advantage 284
competitive behavior 129
  and creativity 27-31, 35-7
compilers 240
Complexity Theory 157
Computer-Aided Design (CAD) 152-3, 314-15
Computer Lib 11

computer manufacture 59, 179
  environmental costs 190-1, 193-5
  human costs 15, 57, 58, 194, 198-9
connection-making 264-6
consumption 123-5, 126, 129
control options 273-4
cookies 221
Cooley, Mike 166, 312
Coopersmith, Jonathan 257, 259
copyright 151, 221
corporations 9, 28, 170, 263
  see also under named corporations
cost-benefit analysis 270
Cottrell, Allin 263
Courbet, Gustave 337
Coyle, Diane 190
craft work 29, 164-5, 317
creativity 27-31, 35-7
credibility 337-9
credit 156, 159
cruelty 50, 82
Cuba 53, 93, 102, 143, 297, 311
currency flotation 55
Cyberfolk 296
cybernetics 54, 265, 268, 271-2, 286, 313
cybersecurity 227
Cybersyn 275, 276, 286-92, 293, 295, 304, 319
cycling 319-20
Czechoslovakia 243

Da Vinci, Leonardo 67
Dahl, Ole-Johan 54
Dantzig, George 261
data centers 213-17
De Gaulle, Charles 48, 52
debt 55, 61, 100, 156, 157, 282
The Defense Calculator 42

Dell 249
Democratic Republic of the Congo (DRC) 198-9
Deng Xiaoping 80, 107, 283
Denmark 95, 121, 143, 172
derivatives trading 157
Diamond, Jared 99
Difference Engine 231, 235
differential analyzers 235, 258
Digital Equipment Corporation 244
digital signal processors (DSPs) 181
Dijkstra, Edsger 242
disasters *see* mutual aid
disruptive innovation 186-7
Dobson, Andrew 114
domination, freedom from 334, 337
Donne, John 330-1
Dorling, Danny 10, 85, 92, 95, 138, 159, 160
dotcom bubble 223-4
Douglass, John David 332
DRC *see* Democratic Republic of the Congo
DSPs *see* digital signal processors
Dutch Telecom 264
Dyer-Witheford, Nick 197, 254
Dykstra 321
Dyson, George 26, 239

Eckert, Presper 40, 41, 155, 231, 238, 239
ecological footprints 93, 95, 135, 212, 303, 306
ecology movement 265
e-commerce 133, 223
economic crises 36, 47, 51, 75, 76, 87, 94, 117-18, 156, 158, 248, 306, 328

economic planning 251-77
e-crime 227
education 49, 79, 93, 94, 136-43, 167, 311, 345
EDVAC 41
Edwards, Paul 42, 177, 270
egalitarian hopes for computing 53-5
egalitarianism 25-7, 68-9, 77-81, 93, 94, 253, 283-6, 311, 336
Egypt 56
e-learning 140-3
electricity consumption 212, 217
electricity distribution 235, 258
electrification, Soviet 257-60
Electronic Associates 237
Electronic Numerical Integrator And Computer *see* ENIAC
electronics industry 33, 58, 107, 131-4, 152, 153, 192
elites 13, 28, 29, 37, 53, 60-1, 62, 65-6, 79, 98, 101, 119, 121, 125, 128, 136, 137, 155, 156, 171, 187, 283, 314, 334, 337
Ellis, Joseph 147
Ellsberg, Daniel 300
email storage *see* cloud computing
Emden, Maarten van 40, 41, 240
emissions, greenhouse-gas 92, 103, 108, 112, 114, 160, 307
employment *see* work
emulative consumption 124-5, 126
Engelbart, Douglas 246
Engels, Frederick 30
ENIAC 39, 40, 155, 231, 238, 239
Enron 119, 158
Enterprise Resource Planning (ERP) 254, 263
entropy 200, 201-3, 206-7, 217, 318

environmental crises 63-4, 299, 303, 306
environmental impact
 of inequality 91-110
 of military activities 93, 162
 of technology 18-19, 30, 56, 111-22, 190-1, 193-5, 208, 212, 217, 239
 of wealth 91-7, 124-5, 137
Epstein, Steven 29
EPZs *see* Export Processing Zones
ERP *see* Enterprise Resource Planning
Espejo, Raul 285, 295, 297
Ethiopia 308
Europe 64-5, 68, 81, 101, 105, 67, 197, 235
 see also under countries
Evans, Christopher 280, 310
Export Processing Zones (EPZs) 107
Extensible Markup Language (XML) 224
'eye-pat' equation 103, 105

fabrication plants (fabs) 179-80, 191, 192, 193, 202
Facebook 18, 186, 196, 214, 215, 216, 304
family honor 336
famines 63, 86, 87, 101-2
Fanon, Frantz 336
Fascism 47, 52
'faster is slower' effect 129-31
Fels, Margaret F 172, 173, 175
Fifth Generation Project 240
financial services sector 48, 134
Finland 76, 94, 95, 167
firearms 60, 65, 67-9
First World War 38, 51, 86, 321
Fisher, Franklin 172

355

Flamm, Kenneth 42
floppy disks 247
Flores, Fernado 286, 288
Floud, Roderick 84
Flowers, TH (Tommy) 39, 40, 238
Foerster, Heinz von 274
food production *see* agriculture
Ford, Henry 78, 174, 252
forests 69, 93, 98, 113, 299, 307
FOSS *see* free and open source
Fourier, Joseph 234
fourth-generation languages 165
Foxconn 15, 58-9, 107
France 8, 47-8, 50, 51, 52, 56, 66, 67, 68, 80, 95, 121, 125, 129, 143, 171, 212, 245, 256, 265, 300, 310, 328, 344
Frank, Robert 127
free and open source (FOSS) 18, 240
free trade agreements 282
Freedland, Jonathan 30
Freeland, Chrystia 136
freight transportation 106, 107, 108
Friedman, Milton 281, 282
future, visioning the 182-5, 303-46

Gaia Hypothesis 303
Galbraith, John Kenneth 31, 156
games, computer 197, 211, 226, 246
Garrett, Jesse James 224
Garton-Ash, Timothy 305
Gates, Bill 12, 146, 155
General Motors 174-5
General Post Office (GPO) 39, 40, 200-1
General Systems Theory 265, 313
generative grammars 313

Germany 32, 48, 52, 59, 60, 62, 95, 130, 212, 320, 345
Gibbons, Steve 137
Gimson, Ernest 309
Gini coefficients 56, 95, 283, 298
Gleick, James 12-13
Global Footprint Network 303
Global Hectares 92-3, 95
Gmail 212-14, 224
GNU/Linux 11, 18, 248
Godowsky, Leopold, Jr 32
GOELRO 257-60
gold standard 76
Goodall, Chris 114
Google 17, 18, 27, 213, 214, 215, 224, 304
Gorbachev, Mikhail 255
Gordon, JE 175, 176, 322
Gordon, Robert 116, 117, 134
governments
    innovation funding 32, 34, 37, 42, 240
Graeber, David 100, 155, 339
graphics 221, 223, 226
grassroots projects 270
great compression 50-3
Greece 95, 248, 256, 345
Green Party 340
Greenbaum, Joan 165
Greenland 215
Greenpeace 196, 212
Gregory, Annabel 84
Grilliches, Zvi 172
Grinevetsky, Vasiliy 258
Grossman, Elizabeth 191, 197
group therapy movement 49
guanxi 309
Gutenberg, Johannes 62
Gutowski, Timothy 200, 208

Haber process 86
hacking 11, 245

Index

Hammond, Barbara 149
Hammond, John 149
Harburg, Yip 51
Harris, Nigel 30
Harvey, David 30
Haslam, Jonathan 299
Hayter, Teresa 101
Hayter, William 83
HCI *see* Human-Computer Interface
HDI *see* Human Development Index
health consequences of inequality 73-90
height loss 83-6, 98
Helbing, Dirk 130
Herlich, Paul 103
Hessel, Stéphane 47
hierarchies 69, 286
  absence of 327
  and mindset 36-7
  see also corporations *and* elites
Hightower, Jim 191, 197
Hildyard, Nicholas 86
Hirsch, Fred 126, 127, 128, 136
history of computers
  (1830s) 20, 38
  (1900s) 38
  (1920s) 258
  (1940s) 39-40
  (1950s) 42, 242-3
  (1960s) 244, 245
  (1970s) 244
  (1980s) 245-6
  (1990s) 240
Hitler, Adolf 52
Hoare, CAR (Charlie) 241
Hobbes, Thomas 304
Hodges, Andrew 40
Hodgson, Geoffrey 262
Hollande, François 344
Holy See 95

Homebrew Computer Club 12
homelessness 162
homeostasis 272
Honeywell 237
Hong Kong 53
housing 69, 105-6, 128, 137, 274, 284
Hsiao-Hung Pai 197
HTML *see* HyperText Markup Language
Huhne, Chris 114, 202
Human-Computer Interface (HCI) 269
human costs of technology 15, 57, 58, 194, 198-9
Human Development Index (HDI) 93
human nature 23-4, 127-9
Hungary 41, 257
Hurricane Katrina 305
Huws, Ursula 141, 163, 324, 325
Huxley, Thomas 329
hypertext 12, 219
HyperText Markup Language (HTML) 219, 221, 224

IAS Machine (Johnniac) 239
IBM Corporation 33, 42, 157, 214, 242, 244, 245
Iceland 95, 214
IfE *see* Initiative for Equality
IMF *see* International Monetary Fund
immigration controls 345
income and wealth gaps 17, 69, 50-3, 55-7, 79, 156, 306, 340
India 29, 56, 60, 68, 96, 104, 108, 121, 154, 197, 212, 256, 286
  see also Ladakh
indigenous peoples 84, 94, 25-6, 283, 310, 311, 312, 336, 337
Indonesia 53, 59, 153

Industrial Revolution 149
inequalities
   continuation of 9-10
   counter measures 339-40
   environmental costs 91-110
   future effects 86-8
   health effects 73-90
   increase in 65
   justification of 129, 342-5
   reduction in 50-3, 65
   and technology 46-72
   tolerance of 13, 140, 337
   see also Gini coefficients
INFOMED 143
Initiative for Equality (IfE) 64
Inmos Transputer 240
innovation 28
   and capitalism 31-3, 35-7
   disruptive 186-7
   public funding 32, 34, 37, 42, 118, 240
   suppression 154-5, 243
input-output analysis 260, 261, 262, 263, 265
Institute for Economic Democracy 255
Intel Corporation 179, 192, 240
intellectual property 12-13, 17-18, 146, 148, 150-5, 221
   see also copyright; patents and trademarks
International Monetary Fund (IMF) 282
International Panel on Climate Change (IPCC) 97
internet 55, 135, 212, 217-18, 326
I=PAT equation 103, 105
IPCC see International Panel on Climate Change
Iraq 97
Ireland 38, 68, 95, 102
Israel 282

Italy 29, 56, 58, 60, 61, 64, 68, 82, 95, 328

Jacobson, Pauline 331
Japan 8, 51, 53, 54, 63, 94, 105, 130, 212, 221, 240, 247, 300
   Tokugawa period 64, 68-9, 97, 102
Jara, Victor, 296
Jarzombek, Mark 312
Java language 222
Javascript 222
Jenkins, Robin 148
Jevons, William Stanley 18, 111, 123
Jevons Paradox 18, 111, 112, 200, 222
Jewkes, John 31, 32
Jing Chen 318
jobs see work
Jobs, Steve 58
Johnniac see IAS Machine
Jones, Martin 25, 84, 98, 99
Joseph Rowntree Foundation 160
justification of inequality 129, 342-5

Kahneman, Daniel 36, 120
Kanotorovich, Leonid 261
Kaser, Michael 255, 259
Kay, Alan 246
Kelly, Kevin 186
Kelvin, Lord see Thompson, William
Kempf, Hervé 125
Kenya 127
Kevlar 177
Keynes, John Maynard 232, 310
Keysen, Karl 172
Khrushchev, Nikita 79, 254
King, FH 63, 308
King's Fund 75
Kissinger, Henry 281

Index

Klein, Naomi 281
knowledge
  and market economy 59-62
  sharing 25, 143, 218
knowledge 'ownership' *see* intellectual property
Knuth, Donald 155
Kodachrome 32
Koopmans, Tjalling 261
Kraft, Phil 165
Kropotkin, Petr 77, 147, 149, 329, 334
Krugman, Paul 55
Kuehr, Ruediger 190
Kurzweil, Ray 112, 238
Kushner, Boris 258-9
Kuznets, Simon 113
Kuznets Curve 113-15
Kyoto Agreement 108

Ladakh 67, 99, 207
Lakner, Christopher 57
Lakoff, George 343-4, 345
land ownership 81, 147-8, 283, 307, 345
Lange, Oskar 256
laser technology 237
Laval, Pierre 52
Lebanon 75
Lebedev, Sergei Alexeyevich 244
LED lights 114, 194, 202
Lee, Richard 310
legacy databases 226
Lenin, Vladimir 252, 258
Leontief, Wassily 260
Lepore, Jill 186
Leroy-Beaulieu, Paul 129
Lesk, Michael 210
Levine, David K 154
Liechtenstein 95
life expectancy 69, 75-7
lifelong learning 140-1

Lindemann, Frederick 268
linear programming 261-2, 263-4
Linux 11, 18, 248
Lis, Katharina 63
living standards 126
Logo language 140
Lohman, Larry 86
Lovelace, Ada 20, 38
Lovelock, James 303
low-impact lifestyles 105-6
Ludgate, Percy 38
Lushai Hills effect 104
Luther, Martin 62
Luxembourg 95
Lynas 195

MacArthur, Douglas 51
Machin, Stephen 137
MacKenzie, Donald 159, 179, 237
MacQuillan, Jim 248
Maisky, Ivan 52
Malaysia 53, 153, 195
Malta 95
Malthus, Thomas 99, 101
Mandelbröt, Benoit 157, 317
Mannes, Leopold 32
Mao Zedong 87
Marglin, Stephen 201, 306, 325
Marine Society 85
market economy 59-62
markup language 219, 220
Marmot, Michael 81
Marx, Karl 30-1, 150
Marxism 30-1
materials, manufacturing 175-7
Maturana, Humberto 287
Mauchly, John 40, 41, 155, 231, 238, 239
Maxwell, James Clerk 203, 204
Maxwell's Demon 204-5
Mayan empire 97-8

Mazurek, Jan 191, 197
Mazzucato, Mariana 37
McChesney, Robert 276
McGilchrist, Iain 36
McGovern, George 50
McNamara, Robert 270
Mead, Margaret 272
media industry 125, 151, 152, 276
medieval societies
  craftworkers 29
  wealth inequality 56, 57-8
Medina, Eden 288, 290, 295
Mediterranean societies, early
  84, 97, 98, 99
memex 218, 235
memory, computer 210-11
Mendel, Gregor 218
Mendes, Chico 125
Merleau-Ponty, Maurice 287
Merton, Robert 158
metals consumption 201
metanorm 336
Mexico 59
microchips 55, 240
  commodification 178-82
  manufacture 190-5
  and obsolescence 131-4
  recyclability 2002
Micropolis 187
microprocessors
  life-cycle costs 202-3
Microsoft Corporation 12, 17,
  155, 214, 215, 217, 221, 224,
  240, 246
Mielants, Eric 61
Mies, Maria 60
migration 68, 82, 101
Mihaileanu, Radu 104
Milanovic, Branko 16, 57
Milgram, Stanley 50
military activities
  environmental impact 93, 162

military spending 161-2, 255
millenarian apocalyptism 115
Miller, Jonathan 198
mining 48, 106, 285, 332
MiniScribe 187
Minitel 8, 245
'Minute Man' 177
misuse of technology 9
mobile phones 15, 37, 42, 57, 58,
  127, 180, 227, 248, 264
Monaco 95
Monbiot, George 306
Monetary National Income
  Automatic Computer
  (MONIAC) 232-3, 234
Montagu, Ashley 316
Moore, Gordon 179
Moore's Law 179, 184, 191, 240,
  242, 271
Moran, James 36
More, Thomas 148
Morocco 104
Morris, William 35
Mosaic 221
Moulin, Jean 52
Mowery, David 34
Mueller, Tobin James 327
Mumford, Enid 269
mutual aid 305, 327-31

Nader, Ralph 172
Napoletano, Janet 227
National Health Service 76, 119
nationalization, industrial 48, 76,
  285, 286, 345
Nazism 47, 52
negentropy 207, 294
Negroponte, Nicholas 184-5
Nelson, Ted 11-12, 219, 220, 221,
  234
Neocleous, Mark 30
neoliberalism 50, 97, 121

Nepal 248, 308
Nest, Michael 198
Netherlands 40, 56, 60, 68, 87, 95, 216, 261, 320, 321
Netscape 224
Netscape Navigator 221
New Zealand 172, 232
Newton, Isaac 24
Nigeria 108
Nike 152, 171
Nixon, Richard 50, 55, 281, 288, 300
Noble, David 142
Norberg-Hodge, Helena 67, 99, 207
Nordhaus, William 113
North Korea 102
Norway 53-4, 95, 103, 143, 172, 311
Nove, Alec 258, 262, 299
nuclear industry 192, 217, 234
nuclear missiles 177, 235, 237, 249
Nygaard, Kristen 53-4, 311, 312

O2 127
Object-Oriented Programming 54
obsolescence 57, 118, 180
 cars 132, 172, 173, 174-5
 computers 132, 195, 196
 microchips 131-4
oil industry 261
oil prices 282
oil transportation 108
old people 48, 81, 310, 313
online shopping *see* e-commerce
Open University 140
operating systems 11, 18, 221, 246, 248
Operational Research (OR) 265, 266-72, 286
Opsroom 291, 292-3, 295
OR *see* Operational Research

Orlov, Yuri 79
Orwell, George 337
outsourcing 57, 68, 119, 161, 184
Owen, Frank 52

Packard, Vance 172
Packer, George 17, 186, 304
Papert, Seymour 140
parenting 49-50
Paris Commune 337
Parra, Angel 289
participatory design 54, 265
patents 22, 29, 41, 147, 151, 152, 154, 155
peasants 56, 63-4, 66-8, 69, 99, 103, 147, 148, 149, 285, 307-9
PERI *see* Political Economy Research Institute
Perry, Grayson 312
Persian empire 68
personal computers
 early history 11, 42, 245
 obsolescence 132
Peru 331
Pettengill, John 67, 68
Pettifor, Ann 100, 101, 155, 339
Peuter, Greg de 197
Pfaff, Donald 330, 331
pharmaceutical industry 80, 152, 154
Phillips, Bill 232
phones *see* mobile phones *and* telephones
photographic industry 32
Pickett, Kate 82, 172, 341
Piketty, Thomas 48, 79, 80, 100, 113, 119, 122, 124, 129, 146, 172, 342
Pinochet, Augusto 281, 297
planning, economic *see* economic planning *and* Cybersyn

Poland 48, 256
Political Economy Research Institute (PERI) 96
Pollitt, Brain 93
population
  affluence and technology 103-4
  and agriculture 98-9, 101-3, 149
pornography sites 214
Porten, Jeff 227, 228
Portugal 95, 148
positional competition 174
positional consumption 129
positionality
  and commodification 149-50
  and human nature 127-9
Pratto, Felicia 82
Preis, Tobias 157
Prince, George 35
prison population 76
privatization 80, 119, 138, 264, 282, 345
programming 17, 28, 155, 164-5, 166, 195, 241
programming languages 12, 54, 140, 165, 222, 240, 241, 311
public asset auctions 264
publishing industry 219, 224
Pye, David 166, 167, 312, 317

queuing theory 256
Quicksort 241-2

racial inequality 9, 47, 76, 87
Rackspace 214
radioactive waste 194
railways 261
RAND Institute 270
Randell, Brian 38, 40
rare earth elements 193-5
Raskin, Jef 247
Ravven, Heidi 27, 82, 329, 330

ray-tracing 314
Reagan, Ronald 282, 283
rebound effect *see* Jevons Paradox
reciprocity cultures 309
recyclability 200
recycling of e-waste 196, 197
regulation
  deregulation 134
  exemptions 192
relocation 135, 153
Renault, Louis 48
Requisite Variety, Law of 273-4, 276
resistance movements 47-8, 52, 256
Revkin, Andy 97
Ridley, Matt 205, 206
risk management 166, 167
Rivoli, Pietra 59
roads 318
Robinson, Joan 232
Rodolfo Debenedetti Foundation 34
Rogers, Deborah 64
Rojas, Alejandro 299
Rooke, Rob 51
Roosevelt, Franklin D 51
Root, Amanda 135
Rosenberg, Nathan 34, 180
Rosenbleuth, Arturo 272
Rosenbrock, Howard 104, 166, 312
Rudgley, Richard 24
Ruskin, John 333
Russia 62, 147, 161, 212, 231, 241
  computer development 242-4
  see also USSR

Sahlberg, Pasi 94
Sahlins, Marshall 28
Saito, Osamu 69

Index

sales effort 170-89
Samsung 179
San Marino 95
Saudi Arabia 161
Saunders, Peter 341, 342, 343
Sawers, David 32
Scandinavia 8, 53, 87, 94, 105, 265
  see also Denmark; Norway and Sweden; for other Nordic countries see Finland and Iceland
Schatsberg, Eric 177
Schiffrin, André 276
Schmidt, Conrad 320
Schmitt, John 141
Scholes, Myron 157, 158
Schön, Donald 31
Schor, Juliet 111, 113
Schrödinger, Erwin 207
Seagate 187
Second World War 32, 37, 39-40, 49, 50, 52, 73, 76, 78, 87, 175, 213, 218, 235, 237, 242, 256, 260, 265, 266-7, 271, 300, 311, 321, 335
Sekula, Allan 107
Semkov, BF 262
Sen, Amartya 102
Setun 231, 243
Sever, Thomas 98
SEZs see Special Economic Zones
SGML see Standard Generalized Markup Language
Shell, Donald 241
ships 29, 60, 65, 261, 320, 321
  containerized 107-8, 195, 196
  speed 129-30
  wartime losses 266-7
shoe industry 107, 152
Sidanius, Mark 82
silicon 177, 193
Silva, Gustavo 294
Simon, Herbert 252

Simstar 237
SIMULA 54, 311
Singapore 53
SIPRI see Stockholm International Peace Research Institute
skilled work 141, 176, 311
  elimination 132, 162-4
  expansion 312
  over-skilling 139
  see also craft work
Small, James 232, 236, 237
Smith, Adam 123
Smith, Anthony 322
Smith, John 84
Smith, JW 255
Smithsonian Institution 38
Snowden, Edward 118, 213
social media 18, 55, 224, 326
Soddy, Frederick 100, 101
solidarity 27, 81, 82, 253, 255, 256, 329, 331-4, 337
Solnit, Rebecca 327-8, 329, 331, 334
Solow, Robert 116, 117
Solow's Paradox 117
Soly, Hugo 63
Solzhenitsyn, Aleksandr 78
South Africa 248
South Korea 53, 59, 63, 248
Soviet Union see USSR
Spain 52, 95, 248, 345
sparse matrix 263
Special Economic Zones (SEZs) 107, 125, 191, 202
speed see 'faster is slower' effect
speeds, clock 238, 244
The Spirit Level 120, 341, 343
spreadsheets 42-3, 147, 263
Spufford, Peter 61
Sri Lanka 283

Stalin, Josef 52, 78, 86, 253
Stallman, Richard 17-18
Standard Generalized Markup Language (SGML) 219
start-up companies 33, 186
statistical filtering 290, 295
Steckel, Richard 84
steel 173-5
steel industry 32, 286
Stewart, Paul 175, 264
Stillerman, Richard 33
Stites, Richard 77, 252, 253, 255
Stockholm International Peace Research Institute (SIPRI) 161
Stonier, Tom 205
storage *see* memory, computer *and* cloud computing
structural adjustment 282
structured programming 240
Sturt, George 174, 309
suicides 15, 58, 74, 75
Summers, Larry 115
Sun Systems 222
suppression of technology 22, 154, 231-3, 236-7, 243
surveillance 213
Sweden 54, 95, 121, 143, 215
Sweezy, Paul M 170, 171, 172, 187
Switzerland 95, 241
Synectics 35
Syria 97
Systems Analysis 270
Szilard, Leo 204

Taiwan 53, 59
tantalum 112, 198-9
Tanzania 107, 308
taxation 50, 61, 64, 68, 69, 76, 96, 107, 118, 121, 124, 129, 195, 233
taxi drivers 138, 139
Taylor, Frederick Winslow 78, 163, 164, 252, 270

technology
  definition 19, 24
  early history 24-5
  evolution 19-23
  and inequality 46-72
  use of term 11-12
telecottage movement 54
telephones 200-1
telex networks 291, 293, 295
ternary systems 231, 243
Texas Instruments 181
texture 316-18
Thailand 53, 153
Thatcher, Margaret 74, 283, 331
Thermodynamics, Second Law of 203, 206
Thompson, William 234, 235, 236
Thornburg, Kent 87
time-sharing 244-9, 248
tobacco industry 343
Torvalds, Linus 18
toxic waste 194-5
trade unions 48, 51, 54, 107, 141, 331-2, 336, 344
trademarks 151
traffic waves 130-1
training 183
transclusion 221
Tressell, Robert 311
TRIPS 153
Trist, Eric 269
Tropp, Henry 38
trust 161, 216, 253, 327, 329, 339
truth-telling 337
Turing, Alan 39, 41
Turley, Jim 191
Turner, Fred 185
Turner, John Charlewood 274
Tymshare 245

Uexküll, Jakob von 266

UK *see* Britain
Ukraine 262
ultrastability 273
unequal societies *see* inequalities
United Nations 197, 256
United States 8, 9, 32, 50, 51, 59, 68, 76, 82, 93, 105, 108, 114, 117, 118, 119, 124, 136, 147, 154, 158, 172, 192, 211, 212, 213, 215, 221, 223, 227, 231, 243, 251, 261, 265, 270, 271, 282, 283, 293, 300, 320, 325
  defense 42, 161, 177-8
  environmental impacts 96, 191, 196-7
  health 75, 81, 85-6, 143, 196
  prisons 76, 160-1
  technology development (1920s) 258; (1950s) 42; (1960s) 31, 245; (1970s) 11-12; (1980s) 245-6
  Vietnam War 55, 270, 282, 300
  wealth inequality 79, 118, 156, 283, 342
Universal Declaration of Human Rights 47, 346
Unix 18, 248
user interface 246-7
USSR 31-2, 51-2, 77-80, 93, 102, 143, 177, 235, 251, 261, 282
  centralized planning 251-62
  electrification 257-60
  health 77, 86-7
  see also Russia
Utopia, building 303-46

value measurements 199-201, 202
Varela, Fransisco 287
variety engineering 275-7
Veblen, Thorstein 125
Venezuela 108

Very Large Scale Integration (VLSI) 180
Via Campesina 308
Vietnam 153, 270, 282, 300
Vietnam War 55
Virtual Reality 240
Visicalc 43

visioning the future 182-5, 303-46
visually-impaired people 226
VLSI *see* Very Large Scale Integration
Von Neumann, John 41, 155, 165, 238, 239
Von Neumann architecture 41-2, 43, 239
Vuskovic, Pedro 297

Wachter, Kenneth 84
Wal-Mart 254, 263
Wallerstein, Immanuel 265
War on Terror 227
Ward, Colin 335
wars
  First World War 38, 51, 86, 321
  Second World War 32, 37, 39-40, 49, 50, 52, 73, 76, 78, 87, 175, 213, 218, 235, 237, 242, 256, 260, 265, 266-7, 271, 300, 311, 321, 335
  Vietnam 55
  Yom Kippur 282
Washington, George 147
water mills 66-7
water supplies 96, 104
Watson, Arthur K 33
Watson, Thomas J 33
Watt, James 111, 204
wealth
  accumulation 79-80, 124
  data 121

disapproval of 335, 336-7
environmental impact 91-7, 124-5, 137
redistribution 281, 341
wealth inequality *see* Gini coefficient *and* income and wealth gaps
Web 2.0 216, 218, 224, 226
weightless economy 190, 191, 205, 239
Wells, HG 310
White, Lynn Townsend 29
Whitehall Studies 81
Wiener, Norbert 54, 268, 271, 286
Wilkinson, Richard 69, 73, 74, 81, 82, 87, 172, 341
Williams, Eric 190, 202, 265
Wilson, Amrit 106
wind power 320
Windows 221, 246, 248
Winnicott, Donald 49
Witt, Robert 83
women 49, 57, 60, 81, 83, 107, 135, 153, 163, 195, 284, 323-4, 325, 328, 336
wood 173-6
Woolgar, Steve 182
work
　autonomy at 81-2, 94, 142, 165-6, 254, 269, 290
　commodification of 107, 150, 161, 162-4
　shared 325-7
　skilled 132, 139, 141, 162-4, 176, 311, 312
　unpaid 310, 323-5

worker suicides 15, 58
working hours 138, 280, 285, 310
World Bank 16, 53, 115, 127, 282, 284, 298
World Resources Institute (WRI) 201
World Trade Organization (WTO) 97, 151-2, 153, 282
World Wide Fund for Nature (WWF) 92, 94, 303
World Wide Web 34, 55, 128, 218
World Wide Web Consortium (W3C) 226
WRI *see* World Resources Institute
Wright, Orville 32
W3C *see* World Wide Web Consortium
WTO *see* World Trade Organization
Wurman, Richard Saul 185
WWF *see* World Wide Fund for Nature

Xanadu 221
Xerox Corporation 33, 246
XML *see* Extensible Markup Language

Yom Kippur war 282

Zachary, Pascal 236
Zapf, Klaus 345
zero-hours contracts 161, 231
Zimbabwe 308
Zizek, Slavoj 115
Zuckerman, Solly 267